D1417857

DISCOVERING THE SECRETS OF THE SUN

DISCOVERING THE SECRETS OF THE SUN

Rudolf Kippenhahn

Translated by
Storm Dunlop

JOHN WILEY & SONS

CHICHESTER · NEW YORK · BRISBANE · TORONTO · SINGAPORE

Other Wiley Editorial Offices

John Wiley & Sons, Inc., 605 Third Avenue,
New York, NY 10158-0012, USA

Jacaranda Wiley Ltd, 33 Park Road, Milton,
Queensland 4064, Australia

John Wiley & Sons (Canada) Ltd, 22 Worcester Road,
Rexdale, Ontario M9W 1L1, Canada

John Wiley & Sons (SEA) Pte Ltd, 37 Jalan Pemimpin #05-04,
Block B, Union Industrial Building, Singapore 2057

A catalogue record for this book is available from the British Library.

ISBN 0 471 94160 3 (cloth)
ISBN 0 471 94363 0 (paper)

Typeset in 10/12pt Palatino from author's disks by Text Processing Deptartment,
John Wiley & Sons Ltd, Chichester
Printed in Great Britain by Biddles Ltd, Guildford and King's Lynn

CONTENTS

PREFACE

This is the fourth of my books on popular science, in which I have tried to introduce ordinary readers to some of the many issues of astronomical research. Starting with the thousands of millions of stars in our Galaxy, I have carried my readers to the edge of the known universe, then back to the strange worlds of the planets, and now to the Sun, our parent star, on (as one might say), a trip round the universe and back. Each of these themes originated in a series of general-studies lectures that I gave to students at the Ludwig-Maximillians University in Munich. I took the Sun as my subject during the summer term of 1987.

I feel that in the ten years since I began, I have learned a lot about the art of writing, and hope that in the future I shall have enough time to come somewhat closer to the ideal. As far as this book is concerned, I have tried to introduce my readers, not only to the Sun's visible phenomena, but also to the basic ideas of plasma physics. So far as I know, these have never appeared in a book intended for a general readership. Once again, I have made use of 'dreams' by my fictional Herr Meyer, with whom my earlier readers will already be familiar. I have also tried to explain in a generally understandable fashion, the question of solar oscillations, because as research makes new discoveries, so it is incumbent on writers of non-fiction to clarify their significance. New scientific results require new methods of explanation, however, and I try to solve this problem in Appendix B, by making use of the idea of a beam of sound.

There are many individuals that I must thank. Part of the text was written at the Kiepenheuer Institute for Solar Physics in Freiburg, to which I had retreated. Roland Buda, Ernst Fürst, Peter Kafka, Helmold Schleicher, Hermann Ulrich Schmidt, Henk Spruit, and Richard Wielebinski have critically examined individual chapters. Alvo von Alvensleben, Wolfgang Duschl, and Hans-Heinrich Voigt have read through the entire text, and helped me with knotty problems. Thomas D. Duval jr., Herbert Friedmann, Sieglinde Hammerschmidt, William C. Livingston, Walter Stein, Alan Title, and Hubertus Wöhl made material available for me. Walter Stein has asked me to mention that the radiotelescope on the roof of the St Michael

Gymnasium in Bad Münstereifel would not have been possible without the co-operation of 'professionals'—the astronomers Dr Wohlleben and Professor Fürst.

The majority of the illustrations were drawn by Jutta Winter, with the line drawings of the circus performances and Figure 5.1 being provided by Evi Kippenhahn. Cornelia Rickl helped with my correspondence, printing out my diskettes, and with the index for the German edition. I also have to thank my wife, Johanna, who not only encouraged me, but also frequently criticized my work. Both were essential in bringing the book to fruition.

I thank Storm Dunlop who has translated the book into English. I have again enjoyed co-operating with him. I also want to thank the personnel of John Wiley & Sons for preparing the English edition.

Göttingen, September 1993 Rudolf Kippenhahn

INTRODUCTION

No one will ever know why the car broke through the parapet of the bridge. As the driver, Godfried Bühren, a patent agent from Osnabrück in Germany, tried to swim to safety, he had a heart attack. The evening news bulletins described the accident because he was involved in an unusual court case, which had been followed with interest by the media. At issue was whether three renowned German astronomers could prove that the interior of the Sun was hot, rather than cold.

This was in the early 1950s. Bühren thought he could prove that only the atmosphere of the Sun was hot. Beneath it the body of the Sun was cold. This could be seen occasionally through the hot clouds covering the surface. He said that the dark spots on the Sun, the sunspots, which have been known since the seventeenth century, were areas where we were looking through holes in the bright cloud-layer and seeing the darker, cooler body underneath.

This idea is contrary to even the most elementary laws of physics. There is, however, nothing unusual in laymen thinking that they have solved some riddle of the universe. Scientific institutions repeatedly receive letters from outsiders, who think they can prove that the experts are wrong, and that, for example, relativity theory is wrong, or that they have invented yet another perpetual-motion machine, even though physics maintains that no such thing can exist. In general, most of these letters cannot even be answered, there are so many of them and, in the majority of cases, the authors would be reluctant to abandon their entrenched ideas. Naturally, it is not impossible for an outsider to have some inspiration that has not occurred to the experts, but modern science presupposes such a vast range of knowledge that it is only after thorough study that anyone can hope to contribute to the frontiers of research. In the field of physics and related sciences in the last 130 years, there has not been, to my knowledge, any case where an outsider has overturned the ideas generally accepted and taught at university level.

Godfried Bühren was a wealthy man, however, and so he was able to establish two prizes, each for 25 000 DM (about $15 000). The first was to be awarded to anyone who could disprove Bühren's theory that a hot cloud cover hid a cold, solid, solar interior, which probably supported vegetation. The second prize was for anyone who could actually prove that the temperature of the solar interior was several million degrees. The patent agent had specified a prize committee, consisting of the Nobel laureate Werner Heisenberg, the physics professor Clemens Schäfer, and a Hamburg lawyer. They were to decide whether any of the submitted essays were worthy of a prize.

Professional astronomers could not simply ignore Bühren's theories, because the public had already been given the impression that they did not actually know whether the interior of the Sun was hot or cold. This is where the principal society for German-speaking astronomers, the Astronomische Gesellschaft, became involved. They were interested in the prizes, because by the end of the Second World War the society had lost all its endowments. The then President, Otto Heckmann, Director of the Hamburg-Bergdorf Observatory, reminded members of their obligations towards the society, and advised them not to participate, as individuals, in the contest for what was, apparently, an easily won prize. Instead, he suggested that two respected members, Professor Ludwig Biermann and Professor Heinrich Siedentopf, and himself, should prepare a submission that would comply with the conditions for the first prize. This they did. The prize committee decided in favour of the manuscript submitted by the three authors, that contradicted Bühren's theory.

But Bühren refused to recognize the award, and took the matter to court. The case went through several hearings. The media took a lively interest, and to outsiders it must have seemed as if it were a later version of the famous case against Galileo Galilei. In fact, there was a major difference between the two cases: Galileo was right, but nevertheless recanted; Godfried Bühren was wrong, but refused to admit it. He lost two different hearings. The judges did not, of course, rule on the temperature of the Sun, but on whether the procedure for selecting the prize had been correctly carried out. Before the case came to a final judgement at the Federal Supreme Court in Karlsruhe, the pugnacious patent agent had died. Following the Karlsruhe court's decision, the society received the first of the two prizes. Because it would have required considerable further legal action, it was finally decided not to apply for the second prize.*

Although nowadays it is relatively easy to prove the Osnabrück patent agent wrong, as late as the last century even so respected an astronomer as Sir William Herschel, the discoverer of Uranus, was convinced that the central regions of the Sun were cold, and that only the solar atmosphere was at a high temperature. At that time, however, thermodynamics, the study of

* The sum of money obtained from the prize—later increased by a bequest—was used to set up a fund from which the society awarded grants to young astronomers for travel expenses to study at other institutes. Herr Bühren is owed some thanks for this, even if he tried to avoid paying the prize during his lifetime. Despite this, my attempt to get the fund named after him—regardless of whether his idea of the Sun was right or wrong—was miserably defeated by my colleagues.

heat, was in its infancy, and it was not yet appreciated that a cool body could not be hidden beneath a hotter shell, without it either being heated up itself, or causing the shell to cool to its own temperature.

We now know that when we look at a sunspot we are not looking at a cooler solar surface through a hot, cloudy atmosphere, and that the temperatures in the centre of the Sun are more than two thousand times higher than those of its surface.

THE SUN IS THE SOURCE OF ALL LIFE ON EARTH

The Sun's gravity holds the Earth, the other planets, and innumerable smaller bodies in its grasp. Even though some of them—such as the block of ice that orbits the Sun and is known as Halley's Comet—occasionally travel so far out into space that sunlight takes many hours to reach them, they are inevitably pulled back towards the Sun.

The Earth does not make any such attempt to escape. It follows an almost circular orbit around the Sun, and always remains approximately the same distance from it. Although the Sun is 150 million kilometres away, and its light takes something over 8 minutes (to be more precise, 500 seconds) to reach us, it is still close enough for its heat to enable life to exist on our planet.

For us, the Sun is the most important body in the sky. Until very recently, our sole sources of energy were those that derived from it. Even hydroelectric power comes from the Sun, because it evaporates water vapour from the oceans and carries it over mountains, where it turns back into water and flows back down through the valleys. When we heat our homes with wood—whether it be in the form of freshly cut timber, or as the coal that it became over a period of millions of years—we are making use of the solar energy that has been stored in plants. Even the energy from oil originated in the Sun. Only the energy from our nuclear power stations does not derive from that source. The fuel that they use, such as uranium, contains energy that ultimately derives from the energy embodied in matter when the universe was born.

Mankind soon realized that life depended on the Sun. It repeatedly appears in the myths of early peoples. According to a Mexican saying, our Sun is no less than the fifth, the preceding four having fallen into the sea.* The Maoris recount the tale of Maui, who lived at a time when the Sun travelled across the sky so fast, that no sooner had one got up than it was time to go to bed, leaving no time to catch any fish. Maui tried to induce the Sun to move more slowly.

* There are authors who gain their living by telling their readers that Venus was flung out of Jupiter, or that it is a portion of an earlier planet that has been destroyed. They attempt to obtain proof of this from accounts given in the Bible or old folktales. I wonder why none of them has taken the four Mexican Suns that have fallen into the sea as inspiration, and turned it into yet another pseudo-scientific best-seller?

After two unsuccessful attempts to snare it using ropes, Maui managed to capture it with his sister's magic hair, and forced it to slow down. Ever since then there has been sufficient time between sunrise and sunset to go fishing. As anyone knows who has seen patient anglers sitting on the harbour wall at Poreç in Yugoslavia or on the landing stage at Santa Cruz in California, they need all the time Maui could give them.

THE ASTRONOMERS' SUN

To modern scientists the Sun is no longer a god, but one of the objects that make up the physical world. As such, it, and all its parts, are subject to the laws of physics. The light that it sends us is not very different from that of a candle-flame here on Earth, or from the radiation that is emitted by a piece of glowing iron. We have learned how to interpret the light from the Sun and know that its luminous surface has a temperature of about 5500°C.

It is not only the visible radiation that we receive that tells us something about the Sun. It also emits forms of radiation that we are unable to see. Its ultraviolet radiation is responsible for the sunburn suffered by careless sunbathers; and we feel its infrared radiation as heat on our skin. Its X-rays also reach the Earth, but luckily they are stopped by the uppermost layers of the atmosphere before they cause damage to life on the surface.

During the Second World War it was discovered that the Sun acted as a radio transmitter, bathing the Earth in radio waves. At first the programme that it is broadcasting may not appear very interesting, but the radio emission from the Sun does, in fact, enable us to determine a great deal of information about, for example, the relatively rapid processes occurring in its outer layers.

Although the Sun may appear perfectly peaceful, when we look at it (perhaps on a warm spring day); when seen through a telescope, its surface layers are constantly bubbling and seething. It immediately reminds us of a boiling liquid. Eventually it was discovered that above the luminous, seething surface layers, blobs of gas are hurled out into space by unseen forces. This material flung out from the Sun reaches us and frequently moves out far beyond the Earth's orbit. We know that this flow of gas from the Sun disturbs the Earth's magnetic field, causes aurorae in the upper layers of the atmosphere, and occasionally cripples short-wave radio communications.

The processes occurring within the Sun are followed not only by solar observatories, such as those on the Canary Islands, on Hawaii, or in New Mexico or California, but also by manned and unmanned spacecraft orbiting the Earth. We study its surface with spaceprobes that investigate it from inside the orbit of Mercury, the innermost planet. We also study it from subterranean laboratories that are located far underground beneath mountains, in caverns hewn from the solid rock, where we search for particles that have come from the centre of the Sun.

We are not only concerned with understanding the Sun. It remains our source of life, and we are more closely linked with it than with any other celestial body. In addition, we are also trying to find ways of making solar energy usable directly by capturing it with large mirrors, and manipulating it in other ways. We are even trying to reproduce on Earth the fusion processes that are the Sun's source of energy. These attempts are still in their initial stages, however, and we do not yet know whether we will eventually achieve the desired goal.

We still depend on the Sun, and this will never change, even if we eventually succeed in reproducing the Sun's energy processes in terrestrial power stations, because our food depends upon plants that grow in sunlight. So we should be thankful that the Sun supplies the Earth with the forms of radiation that allow corn to ripen. The terrestrial atmosphere filters out any solar radiation that is harmful to life on Earth.

In fact, the opposite situation is true: Life on Earth has developed in such a way that it is able to make use of the radiation that arrives at the surface of the Earth. It has evolved to suit the combined result of solar radiation and of the filtration that occurs in the Earth's atmosphere. If, in future, the properties of the Earth's atmospheric envelope are altered — by the emission of harmful chemicals, for example — then we will be exposed to forms of solar radiation to which life has not adapted. If the ozone in our atmosphere is destroyed, for example, as we know it is by the fluorine compounds that have been used in vast quantities, the ultraviolet radiation from the Sun will cause a great amount of damage to life on our planet, which has not evolved to deal with it. At present the protective shield of the Earth's atmosphere blocks out the portion of sunlight that is dangerous to us, and our planet's shell of air ensures that the balance between incoming solar energy and the heat reradiated by the Earth creates a climate in which we can live.

The Sun is therefore the star that we have to thank for our very existence, and upon whose radiation we will continue to depend in the future. Is it so surprising then, that ancient people should have addressed prayers to it as a god?

1

THE SUN'S ENERGY

> The greatest mystery, however is to conceive how so enormous a conflagration (if such it be) can be kept up. Every discovery in chemical science here leaves us completely at a loss, or rather, seems to remove farther the prospect of probable explanation.
>
> John Herschel (1792–1871), *Outlines of Astronomy*, 1875 edn.

The Sun is just one of the approximately 1 thousand million stars in our galaxy, the Milky Way. There is nothing that sets it apart from any of the other stars. It is of importance only to us, because it is the star that provides us with life. Together with the Earth, we are bound to the Sun, and since time immemorial it has supplied us with light and heat.

THE POWER OF THE SUN

Outside the Earth's atmosphere at the Earth's distance from the Sun, 1360 joules of energy per second fall on every square metre of a surface perpendicular to the rays of sunlight. One square metre thus receives an amount of energy equivalent to 1.4 kilowatts. One square metre of the Earth's surface, however, receives considerably less. This is partly because some of the energy is absorbed in the atmosphere, and partly because the rays do not always, or everywhere, arrive at right-angles to the surface. Half the time our square metre of surface is in darkness, and when the weather is bad, any sunlight that reaches it is greatly weakened. Clouds reflect sunlight back into space. In central Europe, for example, one square metre receives an average of about 100 watts. Nevertheless, if we wanted to match the amount of solar energy that one square metre receives throughout the year, by burning heating oil, we would be forced to use about 100 litres. We know what the Sun's radiation means for us, but what does it mean for the Sun?

First, we need to know how far away we are from it. It would be easier for the Sun to provide the energy-equivalent of 100 litres of heating oil per square metre per year if it were nearby than if it were farther away. The Sun radiates energy away to space equally in all directions, and more of this is trapped by a surface that is close to the Sun than by one that is farther away (see Figure 1.1). There is a simple rule to express this: If the distance between the source of radiation and the illuminated surface is doubled, the amount of energy captured is one quarter; at three times the distance, only one ninth; and so on.

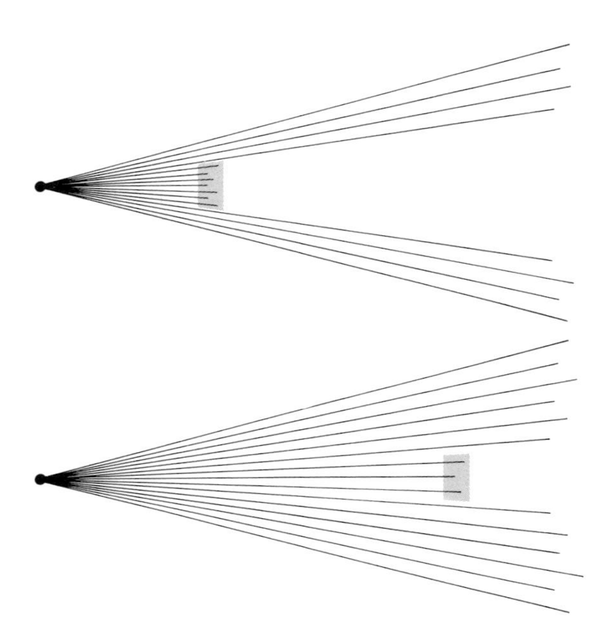

Figure 1.1. Any surface exposed to solar radiation traps energy. If it is close to the Sun (the black dot) as shown above, it intercepts more radiation. A similar-sized surface farther away (below) intercepts far less

How far is the Earth from the Sun? It took a relatively long time for the distance to be determined. Although the Greeks had a very good idea of the distance of the Moon, the distance of the Sun that they determined with a refined, and in principle, correct method, was only about one tenth of the true value. The proper value was first determined in 1672.

In that year Mars came particularly close to the Earth. Observations of the event were made in Paris, and by an expedition to Cayenne, under the French astronomer Jean Richer (1630–1696). If Mars is observed simultaneously from two different sites on Earth, it appears at slightly different positions relative to the fixed, background stars. The smaller the distance between Earth and Mars, the larger the difference in position. This enables the distance of Mars to be determined, when we know the distance between the two observing sites. Shortly before that time, the French astronomer Jean Picard (1620–1682) had measured the radius of the Earth. Once the size of the Earth was known, the distance between Paris and Cayenne could be calculated accurately. This, in turn, enabled the distance between Earth and Mars to be determined.

This was the starting point for distance determinations in the Solar System. If we know the distance between two planetary orbits and also the times that those planets each take to complete one orbit, then the distances of the orbits from the Sun may be calculated. To do so, we use Kepler's Third Law, which uses the observed orbital periods of two planets to determine the ratio between their orbital radii. The orbital periods of planets are easy to determine. As a result, the distance of the Sun was found, in 1672, to be about 150 million kilometres.

Since then we have devised better methods of plumbing distances in the Solar System, and we know the distance exactly: The average distance between the Sun and the Earth is 149 598 000 km. Sunlight takes about eight minutes to cover that distance. If the Sun suddenly switched off one day at 12 noon, we would notice it at 12:08.

Once the distance of the Sun had been determined, its size was also known. In the sky, it appears as a disk with a diameter of half a degree. Simple trigonometry shows that the diameter of the Sun is 1.4 million kilometres. That is about 110 times the diameter of the Earth. If the Earth were situated at the centre of the Sun, the Moon's orbit would still lie inside the Sun. The solar globe would hold more than one million bodies the size of the Earth.

From the amount of radiation falling on our square metre of surface, and from the known distance of the Sun, we can determine the overall amount of radiation produced by the Sun. Expressed in millions of watts (i.e., megawatts) it is a 21-figure number! The amount of radiation poured into empty space in one second would serve the whole of humanity's needs for a million years.

THE SOLAR POWERHOUSE

Any fire goes out when it runs out of fuel, and embers become cold unless we feed them with something. But things seem to be different for the fiery body of the Sun. It provides heat for mankind, and for animals and plants, all of whom use it to live, without cooling down. And it does this, not just day by day and year by year. We now know that it has been radiating an undiminished amount of energy for thousands of millions of years. We need to visualize what a thousand million years means. If we say that a human generation is 30 years, then 66 generations have lived since the time of Christ. (A series of people, one from each generation, from father to grandfather, to great-grandfather, and back to the beginning of our calendar would therefore have ample space in a single railway carriage.) Ignoring the fact that Mankind has appeared late in the history of the Earth, we can calculate the number of generations in one thousand million years. It comes to 30 million, about 30 times as many inhabitants as the city of Munich in Germany, or about four-and-a-half times the population of New York. This is an inconceivably long time, during which the Sun has been radiating away its energy to space at a perfectly steady rate!

Is there some source of power hidden in its interior that we have to thank for this unbelievable amount of energy? This question inevitably arises, because we are familiar with the law of the conservation of energy. Energy can only be derived from another form of energy. Light from a lamp comes from electrical energy, supplied to us by a power plant. The latter in turn uses the principle, which physicists also call the energy law, in its planning: Energy is not lost, and nothing can arise from nothing. In their calculations, electricity companies take account of the fact that none of the energy transmitted over their supply systems is lost, and that it all reaches their customers, who pay for it. In fact, some of the energy is converted into heat on its way to the users, and raises the temperature of the supply cables. Energy is not lost, it may only become unusable.

It is Nature's chameleon. In electrical batteries, electrical energy is produced from chemical energy; in an electric light bulb, electrical energy turns into light and heat. In a cycle dynamo, the mechanical energy derived from the motion of the bicycle is converted into electrical energy. At the same time, however, the cycle's kinetic energy—or energy of motion—is also converted into heat, because the tyres become warm as you ride along. In doing so, the wheels are braked slightly. The law of the conservation of energy, which is so important for physics as a whole, was first discovered in the last century. Because it applies to a whole range of different forms of energy, it was noticed in several fields of physics. The decisive finding was made by the son of a Schwäbian pharmacist, the doctor, Julius Robert Mayer (1814–1878), but the Englishman, James Prescott Joule (1818–1889), and the German doctor and physicist, Hermann von Helmholtz (1821–1894), expressed the law in more precise terms. As people learnt that energy could only be derived from energy, the inevitable question arose as to what was the source of the Sun's energy.

It derives from the material that forms the Sun. Anyone who wants to know what drives the Sun's energy source must first take into account the substances that occur within that gigantic sphere of hot gas. For a long time no one knew what materials existed in the Sun, but as early as the eighteenth century it was possible to determine its overall mass—to weigh the Sun from a distance, so to speak.

Even today, the determination of the mass of a celestial object presents astronomers with some difficulty. We can say something about the mass of a star only if another body is moving in its gravitational field. In the case of our own star, the Sun, the motion of the planets tells us how much mass exists in the Sun. Isaac Newton (1643–1727) established how gravitational attraction decreases as the distance between two bodies increases, but that was not enough to obtain the mass of the Sun from the motion of the planets. It was only when the British scientist Henry Cavendish (1731–1810) succeeded in carrying out a difficult measurement, in the laboratory, of the mutual gravitational attraction between two lead spheres, that the law of gravitation was obtained in its final form. It then became possible to determine the amount of mass in the Sun from the orbits of the planets.

We now know what a vast amount of material is to be found in the Sun. Expressed in tonnes it is a 28-figure number. That is more than 300 000 times the mass of the Earth. With a knowledge of the amount of material in the Sun it became feasible to examine the question of what was the source of the Sun's energy.

Naturally, the idea first considered was some form of combustion, which was the most familiar method of producing considerable heat, and where chemical energy is turned into heat. But the Sun would be unable to sustain its energy requirements for very long using chemical fuels. That much was known to John Herschel (1792–1871), which was why he wrote the words quoted at the beginning of this chapter. He added 'If conjecture might be hazarded, we should look rather to the known possibility of an indefinite generation of heat by friction, or to its excitement by the electric discharge, than to any actual combustion of ponderable fuel, whether solid or gaseous, for the origin of the solar radiation.'

It will be seen that on the one hand he felt that the Sun's energy was derived from another form of energy (and was thus drawing on the law of the conservation of energy), and on the other he spoke about 'the indefinite generation of heat from friction' and (in a later passage) about 'a continual current of electric matter' that was probably 'constantly circulating in the sun's immediate neighbourhood' and caused the solar atmosphere to glow like the aurora. When John Herschel wrote these passages, Mayer had yet to begin his medical studies.

Where, then, does the Sun's energy come from? In 1846, Mayer himself wondered whether it gained its energy from outside. We know that there are innumerable lumps of rock orbiting within the Solar System, examples of which occasionally reach the surface of the Earth and are then studied by scientists, to whom they are known as meteorites, or are exhibited in museums where they are stared at in wonder by the general public as 'stones from space'. If meteorites fall on the Earth, which is small, far more must encounter the Sun, which is gigantic. If one that has a mass of a kilogramme falls into the Sun—its speed is around 600 km / s—approximately 200 thousand million joules are converted into heat. That is more than 55 000 kilowatt-hours. At an electricity price of 0.105 units per kilowatt-hour the amount of energy released would cost 5775.00 units. In principle, the Sun's energy loss could be met by a continuous rain of meteorites, at a rate of two thousand million per second! Tonnes of meteoritic material would have to fall into the Sun. In a year, it would still be a tiny amount when compared with the Sun's total mass, but the planets are sensitive indicators of any possible mass increase. If the Sun's gravity were to increase, the planets would begin to move faster in their orbits. Centuries of observation of the motion of the planets, however, gives no indication of any increase in the mass of the Sun—certainly none as rapid as would be required to explain its radiation through the accretion of meteoritic material.

Hermann von Helmholtz was the first to draw attention to another process capable of providing energy. In a public lecture in 1854, he first discussed the question of whether a chemical process, such as combustion, could provide the energy radiated by the Sun. As we have seen, John Herschel had already described chemical energy as being inadequate. Helmholtz explained that the reaction that released the most energy from the least mass (the combination of hydrogen and oxygen to give water), could meet the Sun's energy requirements for just 3021 years. Even human history showed that the Sun had been shining for longer than that. Geology indicated that the Sun had been shining for millions of years in a way similar to today. Helmholtz drew the conclusion that no chemical reactions could provide the Sun's energy. He then described the arguments that disproved Mayer's meteoritic hypothesis. He then proposed a new source for the Sun's energy, which no one had considered previously. We now call this *gravitational energy*.

Helmholtz began with the idea—which is still valid today—that the Sun formed from diffuse clouds of gas in space. As these gases contracted, they simultaneously heated up. The pressure in the interior of the collapsing cloud rose, however, and forced the contraction to come to a halt. Subsequently, the sphere of gas could contract only at a rate that allowed it to radiate away the energy that it gained through contraction. We still believe that this process plays an important role in the early stages of a star's life. Helmholtz estimated that the Sun had been shining for 22 million years and that it could continue to do so with undiminished strength for a further 17 million years, before other processes arrested its collapse.

In his book about early solar physics, the astronomer Jack Meadows has described how, around the middle of the last century, many astronomers believed that 20 million years could be considered a reasonable age for the Sun—but primarily because the latest estimates of geological time had not reached them.

In fact, for some time, geologists had repeatedly tried to estimate the age of the Earth. How long did it take for a valley to be cut? How long did it require for sedimentary layers at the bottom of the sea to reach their current thickness? Such considerations led to times of a few hundred million years. If the Earth had possessed a climate that allowed water to exist in liquid form for such a long time, then the Sun must have shone for a far longer time than Helmholtz's collapse mechanism would permit. When Helmholtz proposed his energy mechanism, it had already been overtaken. A far more productive source of energy was required.

So what was the reservoir of energy that allowed the Sun to shine with undiminished strength for such unimaginable lengths of time? The question became even more crucial when the French physicist Antoine Becquerel (1852–1908) discovered radioactivity. To understand this we need to say a little about our picture of the structure of atoms.

ATOMS AND THEIR COMPONENT PARTICLES

Every atom has a nucleus, which is surrounded by a shell of electrons. Atomic nuclei themselves consist of protons, positively charged particles, and neutrons, which are electrically neutral. Both types of particle have approximately the same mass. It takes a 24-figure number of neutrons to make up one gramme. The negatively charged electrons are even less massive, amounting to about one two-thousandth of the mass of a proton. There are the same number of electrons spinning around in the shell as there are protons in the nucleus. The charge arising from the positively charged protons in the nucleus is precisely compensated by the negatively charged electrons. From the outside, therefore, an atom carries no electrical charge. All the chemical properties of materials are determined by these electron shells. Atoms therefore react chemically in the same way when they have the same number of electrons in their shells, or—in what amounts to the same thing—when they have the same number of protons in their nuclei. The chemical properties of an atom do not change if the number of neutrons in the nucleus is altered, because this has no effect on the electron shell, which is governed only by the number of protons in the nucleus. All atoms with the same number of protons are therefore considered to be the same element. For example, all atoms with eight protons are forms of oxygen. In nature, the most common oxygen atom also has eight neutrons, and its nucleus therefore contains 16 particles. This form of oxygen is described as ^{16}O, where O is the symbol for oxygen. In contrast, the ^{17}O atom has eight protons and *nine* neutrons. This sort of oxygen also occurs in nature, but not as frequently as ^{16}O. Atoms that differ only in the number of neutrons are known as *isotopes* of an element.

Hydrogen has three isotopes. The normal hydrogen nucleus consists of a single proton. To compensate for this positive charge, a single electron orbits in its immediate vicinity. It is responsible for hydrogen's chemical properties. There is also *heavy hydrogen*, known as *deuterium*. It nucleus contains a neutron in addition to the single proton. For this reason, the mass of the deuterium nucleus is twice that of ordinary hydrogen. Deuterium is very rare in nature. Out of every 10 000 hydrogen atoms, only one or two are deuterium. Because a deuterium nucleus has the same electrical charge as one of normal hydrogen, a single electron is still able to neutralize the positive charge on the nucleus. Normal hydrogen is written as ^{1}H, and the symbol of deuterium is ^{2}H. The symbol for the third hydrogen isotope, *tritium*, is ^{3}H, and here the proton is accompanied by two neutrons.

Hydrogen is the most abundant element in the universe, and helium comes second. In its most common form, ^{4}He, there are two protons and two neutrons in the nucleus. A helium isotope that plays an important role in the Sun's energy process is ^{3}He. Its nucleus consists of two protons and one neutron.

Apart from the three particles that go to form atoms, namely electrons, protons, and neutrons, there are many other particles in nature. Most exist for an extremely short time before they either disintegrate into other types of particle, or combine with other particles. There is, for example, a counterpart of the electron, the *positron*. It has the same mass as the electron, but the opposite charge. If it encounters an electron, both particles disappear in a flash of energy as their mass is converted into radiation.

One particle that still appears strange to physicists is the *neutrino*. It plays a special part in solar physics, and we shall discuss it further in Chapter 9.

RADIOACTIVE CLOCKS

Many atomic nuclei are radioactive, which means that they disintegrate (or 'decay'). One is the uranium isotope ^{238}U, whose nuclei consist of 92 protons and 146 neutrons. Left to itself for 4.5 thousand million years, half of the atoms of ^{238}U will turn into the lead isotope, ^{206}Pb.* A total of 10 protons and 22 neutrons have been ejected from the nucleus. The atom loses an electron from its shell together with every proton.

Because every radioactive material has a characteristic decay period, the latter may be used as a measure of time. Say we begin with 1000 atoms of ^{238}U and wait. When only 500 uranium atoms remain, alongside 500 new atoms of the lead isotope, ^{206}Pb, 4.5 thousand million years have elapsed.

We might be forgiven for believing that age determinations using the radioactive decay of ^{238}U would be perfectly simple. If a sample contains ^{238}U, all we have to do is to determine how much lead ^{206}Pb is also present. We then know how much time elapsed for the lead to form from the uranium. In reality, everything is much more complicated. Some of the ^{206}Pb may have been present in the sample at the beginning, and not been formed by decay of uranium ^{238}U. To avoid this difficulty, we have to resort to other types of radioactive atoms. There are many other types of atom that are suitable for measuring the time required for geological processes.

Radioactive elements are like purpose-made clocks that serve to measure geological time and they provide new opportunities for determining the true ages of various minerals. At the beginning of this century, a mineral from

* The nuclei in this type of lead consist of 82 protons and 124 neutrons.

Mozambique, zircon, which was found to contain traces of the radioactive element thorium, held the age record at 1.5 thousand million years. The Earth was significantly older than people had thought! If the Earth, one of the Sun's companions, was so old, the Sun itself could hardly be younger. There was no way in which Helmholtz's collapse mechanism could have provided the Sun's energy for such a long period of time.

THE ENERGY CONTAINED WITHIN MATTER

While radioactive age determinations caused the problem of the Sun's energy to become even more critical, a new impetus came from a different direction. In 1905, Albert Einstein (1879–1955) put forward his theory of special relativity, in which he established that, in principle, mass and energy are the same, and may be converted into one another. The law of the conservation of energy and the law of the conservation of mass, which had been proposed by Antoine Laurent Lavoisier (1743–1794) in the eighteenth century, could be combined into a single natural law, in which, according to Albert Einstein, energy and mass were the same. He expressed this in quantitative terms in his famous equation $E = mc^2$. This may be taken to mean that a mass m may be turned into energy E. Similarly, energy E may be converted into mass m. In calculating the actual amounts, the mass must be multiplied by the square of the velocity of light c, which is 300 000 000 cm/s. Let us take as an example a gramme of material, i.e., 0.001 kg. The energy to which this corresponds is $E = 0.001 \times 300\,000\,000 \times 300\,000\,000$ kg m^2/s^2. The unit employed here, 'kilogramme per square metre per second squared' is very unusual. Converted to a more useful energy unit, it is equivalent to 25 million kilowatt-hours. Such large quantities of energy are often expressed in the macabre unit of kilotonnes of trinitrotoluene (TNT), or the amount of energy in 1000 tonnes of that particular chemical explosive. If we were able to convert our gramme of material suddenly into energy, we would obtain an explosion equivalent to 20 kilotonnes of TNT. This is the same as the force of the bomb that destroyed Hiroshima on 6 August 1945. In fact, therefore, only about one gramme of material was converted into energy over Hiroshima. That corresponds to about one tenth of the amount of spread that you use on a slice of bread.

It is not, however, as easy to convert mass into energy as Einstein's equation might suggest. No one has yet succeeded (luckily) in converting a larger amount of material into energy such that nothing remains. Even in the Hiroshima bomb, about one kilogramme of nuclear explosive was used in converting about one thousandth of that amount into energy. Only a small fraction of the mass in atomic nuclei may be turned into radiation. Despite this, nuclear fuels are the most powerful sources of energy yet known.

After 1905, the problem of the Sun's energy source appeared to be solved.

Its mass provides the Sun with a gigantic energy reservoir, from which it can easily make up for its losses as radiation. Astrophysicists could breathe again.

THE SUN'S SOURCE OF ATOMIC ENERGY

In nuclear reactions, the particles forming the atomic nuclei are rearranged. Nuclei that consist of many protons and neutrons may decay, and those with lesser numbers of nuclear particles may combine to give larger nuclei.

Atomic nuclei that are less massive than that of iron are generally very stable objects, in which powerful forces bind the protons and neutrons together. If we want to split an atomic nucleus, we have to add energy. Some isotopes require more, and others less. If we split a helium nucleus into its component particles—two protons and two neutrons—we are forced to add energy. This is the so-called *binding energy* of the helium nucleus. If we convert the four particles back into a helium nucleus, we gain that amount of binding energy. In fact, the mass of a helium nucleus is slightly less than the sum of the mass of the individual particles, because it does not contain the energy released when the particles combine. Because energy equals mass, the mass of a helium nucleus is therefore also less.

The Sun consists mainly of hydrogen and helium. This suggests the question of whether perhaps the whole of its radiated energy arises from the fusion of hydrogen nuclei into helium nuclei. The question is, however, not so simple as it might seem at first sight. The individual steps are shown in Figure 1.2. Note that protons (i.e., positively charged nuclear building blocks), turn into neutrons (electrically neutral particles). The positive charge is carried away in the form of a positron, which is soon annihilated when it encounters an electron and both turn into radiation. Overall, four hydrogen nuclei are converted into one helium nucleus. If one gramme of hydrogen is converted into helium, 180 000 kWh of energy is released in the form of radiation. It is true that this is far less than if the whole gramme of hydrogen were converted into radiation, but it will more than suffice to keep the Sun,

---

Figure 1.2. The Sun's source of energy. Two hydrogen nuclei (H) collide (top). A positron (e^+) and a neutrino (N) are produced. The properties of the positron are similar to those of an electron (which is negative), but it has a positive charge. A deuterium nucleus (D) remains. When this collides with a proton, a nucleus of a helium isotope ^3He is produced, which consists of two protons and one neutron. Simultaneously energy is lost in the form of radiation. When two ^3He nuclei collide, a ^4He-nucleus is produced, together with two free protons. Overall, this chain of reactions has fused four hydrogen nuclei into one helium nucleus. The energy that is thus released gives rise to the Sun's radiation. Alongside the reactions shown here, others also occur, but these are not significant in the Sun's energy production (see Figure 9.1)

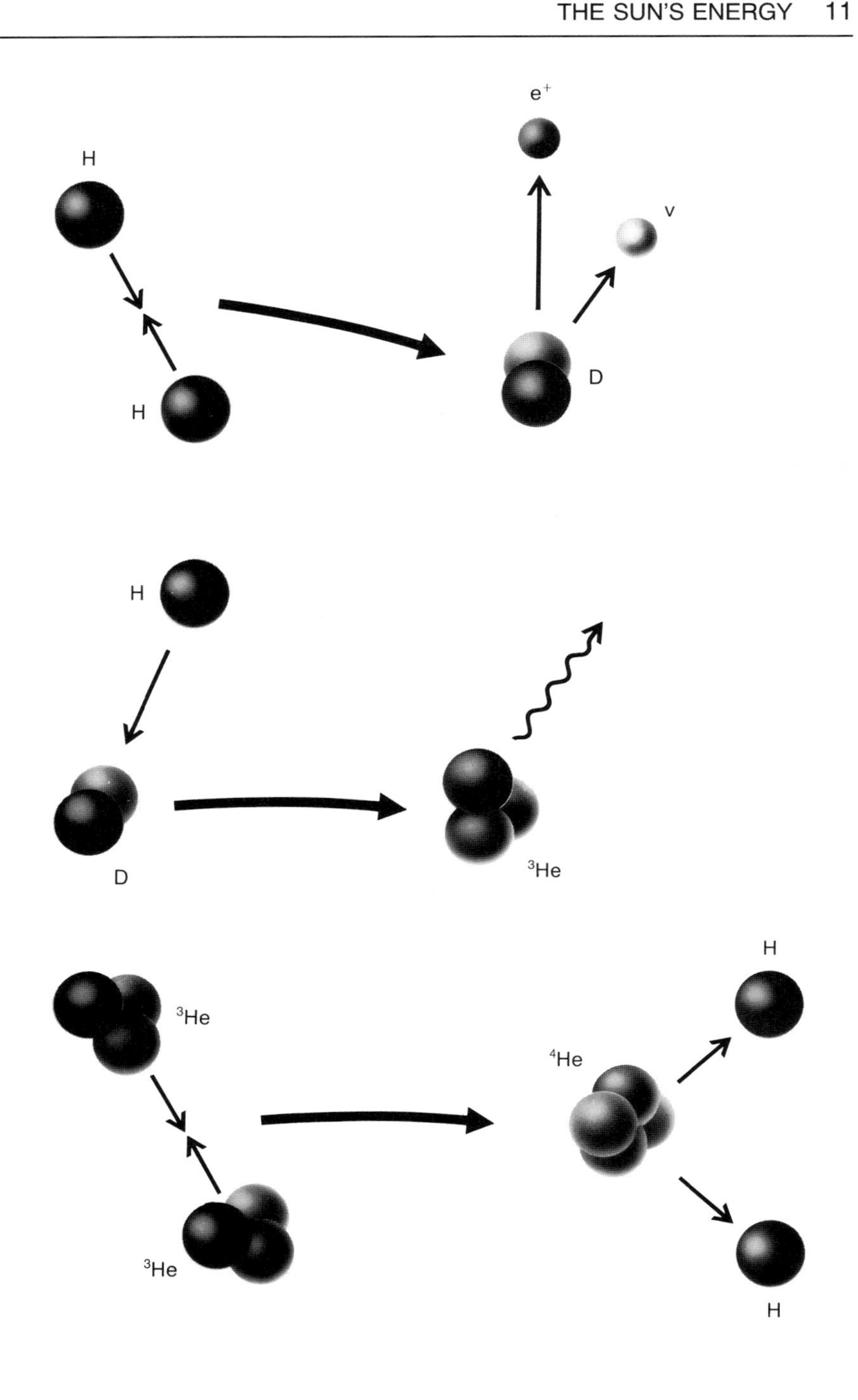
H
H
e+
v
D
H
D
3He
3He
3He
4He
H
H

which originally consisted mainly of hydrogen, shining for 100 thousand million years—significantly longer than the current age of the universe itself.

Einstein's equation implies that as the Sun emits radiation, it also loses mass. Some four million tonnes are lost in the form of radiation every second. Although this may appear a very large amount, it takes 45 million years for a mass equal to that of the Earth to be lost. So even this loss fails to harm the Sun, because its mass is 300 000 times as great.

If we try to estimate the age of the universe, we get, at most, a figure of 20 thousand million years. If the Sun had been shining for the whole of that period with its current luminosity, it would have radiated away one thousandth of its mass.

We have not mentioned the fact that the processes shown in Fig. 1.2 also produce a neutrino. We will return to the matter of this neutrino and of other secondary reactions that are omitted from the diagram, later in Chapter 9.

We now know that the Sun is a gigantic nuclear reactor; or to be precise: a fusion reactor, in which hydrogen is converted into helium. Life on Earth owes its existence to the nuclear reactions shown in Figure 1.2. Attempts are being made on Earth to convert hydrogen into helium as a source of power. The water in the oceans contains more than enough hydrogen. Despite feverish research all over the world, no one has yet succeeded in obtaining a large amount of energy from hydrogen fusion.

The first traces of life on Earth are 3.5 thousand million years old. For at least that length of time, the Earth has been exposed to a steady stream of radiation from the Sun. Life has been able to rely on it. When humans became capable of intellectual thought, it soon became evident that their lives were dependent on the Sun. Is it any wonder that in many cultures, people such as Akhenaton, the Egyptian pharaoh, should have venerated it as a god? When, at the beginning of the seventeenth century, during the Renaissance, it was established that the Sun was not completely free from blemishes, there were those who did not want to believe it was true.

2

SUNSPOTS

> Scheiner was the first who employed blue and green stained glasses in solar observations, which had been proposed seventy years earlier by Apian (Bienewitz), in the Astronomicum Caesarium, and had also been long in use among Belgian pilots. The neglect of this precaution contributed much to Galileo's blindness.
>
> Alexander von Humboldt, *Kosmos* [trans. E.C. Ottë and B.H. Paul, 1871]

We should not be surprised when scientists nowadays argue about who was the first to make a particular discovery. People have always been the same. The history of sunspots is just one among many examples.

WHO WAS FIRST?

Nowadays we award priority in a discovery to whoever first publishes their results. On this basis, sunspots were discovered in East Friesland (in Germany).

David Fabricius was parish priest of Westerhave near Dornum, and later of Osteel in East Friesland. During the day he looked after his congregation, but the night was devoted to the stars. This secondary occupation earned him a reputation that spread far beyond East Friesland. He was in contact with Tycho Brahe, the great Danish astronomer, and he kept up a voluminous correspondence with Johannes Kepler. His eldest son, Johannes, was also drawn to astronomy. However, his father sent him to study medicine at Helmstedt and Wittenberg, and in 1609 to the university of Leiden in Holland. A year before, or perhaps even earlier, the telescope had been invented in Holland. The young Fabricius probably brought home from Leiden the telescope that he later used.

Prompted by his father to study the Sun, on 9 March 1611 he saw, through his telescope, a small dark spot on the disk of the Sun.* At first he thought he was mistaken. He called his father, who also saw the spot. They watched the Sun through the telescope for the rest of the day. The spot did not change. They spent an anxious night, and impatiently waited for sunrise. When the Sun finally rose above the horizon, the spot could still be seen. Johannes Fabricius published news of the discovery in 1611. This established him as having been the first to discover a sunspot.

We do not know exactly when the younger Fabricius died. We know of his death only from an obituary written by Johannes Kepler. The elder Fabricius followed him a short time later. After he had publicly accused a farmer, from the pulpit, of having stolen geese from him, he was killed by a blow from a spade. His son Johannes may have been the first to describe sunspots, but someone else was actually the first to see them.

They had been recorded, using the recently discovered telescope, on 8 December 1610, by the English mathematician and philosopher, Thomas Harriot (1560–1621), although this was admittedly only established from his unpublished papers 200 years later. Three days before Fabricius, namely on 6 March 1611, at Ingolstadt, the Jesuit priest, Christoph Scheiner (1575–1650) and his pupil and assistant, Johann Baptist Cystat (1588–1657), saw spots on the disk of the Sun. However, Scheiner's religious superior would not accept the idea of spots on the Sun. 'Such a thing', he is supposed to have said, 'has never been mentioned by any ancient philosopher: I have read my Aristotle through from beginning to end more than once, and found nothing at all like this. So keep quiet about this absurd idea, and don't make a fool of yourself in public. Instead you should convince yourself that it is simply some fault in your eye or your telescope that makes you think you see spots on the Sun.' (This quotation is attributed to the monk in a book of popular lectures on astronomy, published in 1844 by the high-school teacher Lorenz Wöckel from Nuremberg.) Rather than deliberately disobeying this advice, Scheiner only mentioned the matter in two letters to the Augsburg patrician Marcus Welser (1558–1614). Welser, however, had the letters printed. To avoid Scheiner having problems with his superior, they appeared under the pseudonym of Apelles. When this came to Galileo Galilei's notice, he wrote to Welser in Augsburg and claimed that he had seen spots on the Sun long before, namely in 1610. This began an argument between Scheiner and

* If one reads Johannes Fabricius' original letter, one is horrified at the heedless way in which both father and son looked directly at the Sun through their telescope. That neither of them lost their sight they owe to the extremely low light-grasp of their instrument. It was only later that they used a form of pinhole camera to observe the Sun. The projection method, which Scheiner employed (Figure 2.2) is difficult with the Dutch telescope, which consisted of one convex and one concave lens. At this point, I must strenuously warn readers of the danger of looking at the Sun through a telescope or a pair of binoculars, unless they are prepared to be fully responsible for the fact that someone else will have to read the rest of this book to them. A safe projection method of observing the Sun is described in Appendix A.

Galileo over priority.* Some of Galileo's drawings of sunspots, made in 1612, are shown in Figure 2.1.

Scheiner later lived in Rome for nine years, during which time he published his major astronomical work, *Rosa Ursina,* in 1630. This contained notes of numerous observations of sunspots. By that time, Scheiner had come to be regarded by the Jesuits as a renowned scientist, and so the work subsequently appeared under his own name. He was unable to refrain from numerous sarcastic remarks about Galileo.† Although Galileo and Scheiner may have fought one another bitterly, they were united in ignoring Fabricius' discovery.

In his observations of the Sun, Scheiner used the method shown in Figure 2.2 (top), where the telescope is directed at the Sun, and the light that passes through the tube is allowed to fall on a white screen. If the telescope's optics are properly adjusted, a sharp image of the Sun appears on the screen. Appendix A describes how anyone may use a similar method with a simple pair of field glasses to observe sunspots.

In fact, people should have noticed sunspots long before the telescope was discovered, because whenever rays of sunlight pass through a small chink in a window's shutters, or the crack of a door, and fall on a wall opposite, a tiny image of the Sun's disk appears, in which it is occasionally possible to recognize large sunspots as dark spots. In 1613, the Holy Roman Emperor Matthias travelled to Regensburg where a meeting of the Electors was being held. Because it was planned to introduce the Gregorian calendar, Johannes Kepler, who was then working in Linz, was asked to attend as a 'calendar specialist'. So in July he first visited the town, where he was to be buried 17 years later. During a visit to the cathedral, he noticed that circular images of the Sun were projected onto the floor by sunlight penetrating tiny openings in the roof. (This is similar to the principle on which a pinhole camera operates.) He later recalled 'In the cathedral I saw, and drew to the attention of nearby acquaintances, traces of sunspots in the small circles of light that fell through gaps in the window high above.' It would, of course, have been possible to see this before the discovery of the telescope, because, as mentioned above, it is not essential to have a cathedral to act as a pinhole camera.

From time to time, one of the spots on the Sun reaches a diameter of more than 50 000 km. It is then even possible to see it with the naked eye. About 1 per cent of all sunspots or tightly clustered groups of sunspots reach this size. Naturally, the spot or group of spots may not be seen against the blinding disk of the Sun, but they may be seen when the light from the latter is dimmed, such as by haze or cloud, and it is possible to look at the disk without other

* The rivalry was all the more pronounced, because Scheiner believed in the old, Greek cosmology, where the Sun, the Moon, and the planets all orbited the Earth, whereas Galileo was a passionate supporter of Copernicus, who had taught (as early as 1543) that the Sun lies at the centre of the planetary orbits, and that the Moon alone orbits the Earth, while the Earth and all the other planets orbit the Sun.

† The claim that Scheiner played a part in instigating the trial of Galileo in 1633 cannot be substantiated.

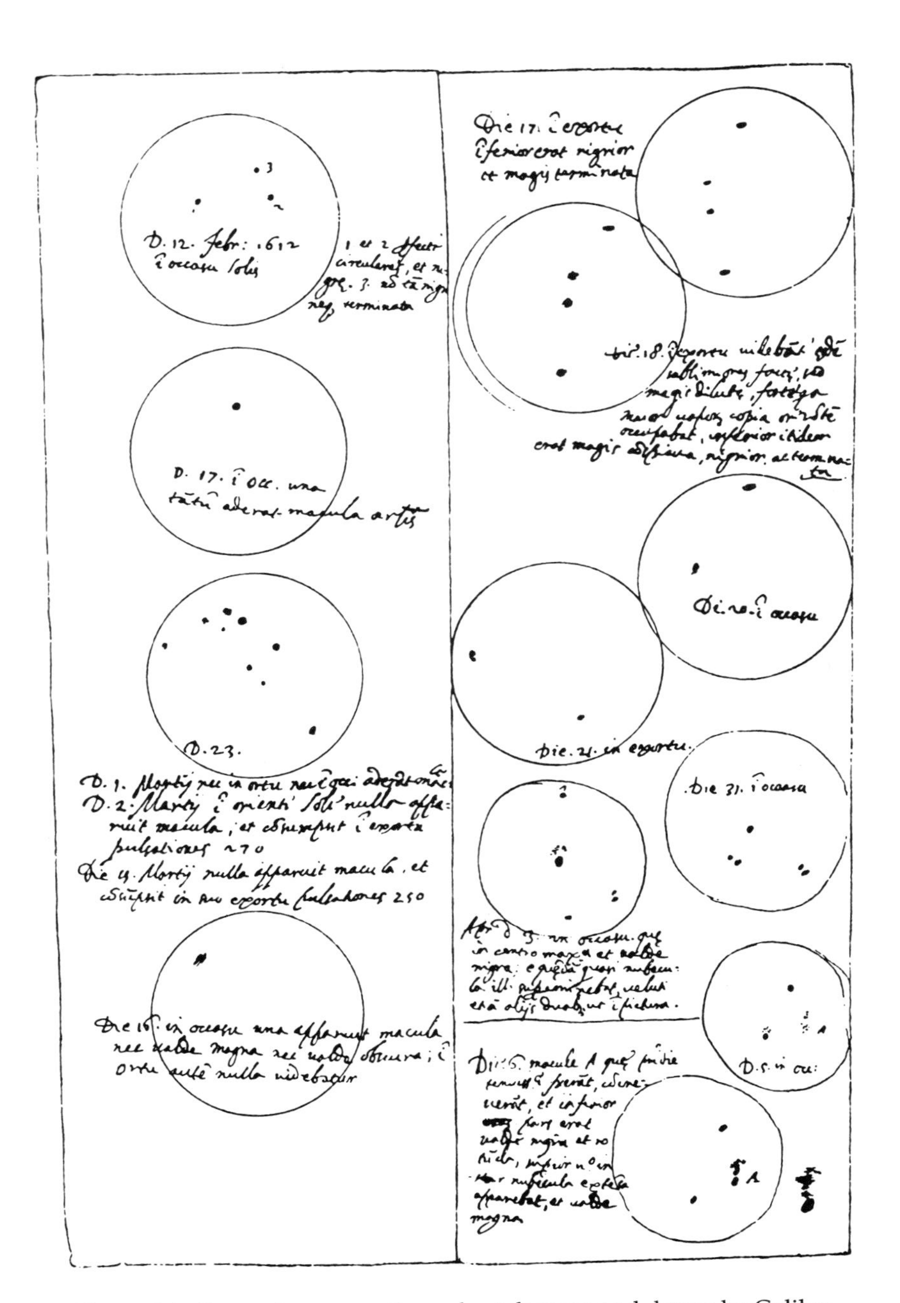

Figure 2.1. Sunspots, as seen through a telescope and drawn by Galileo

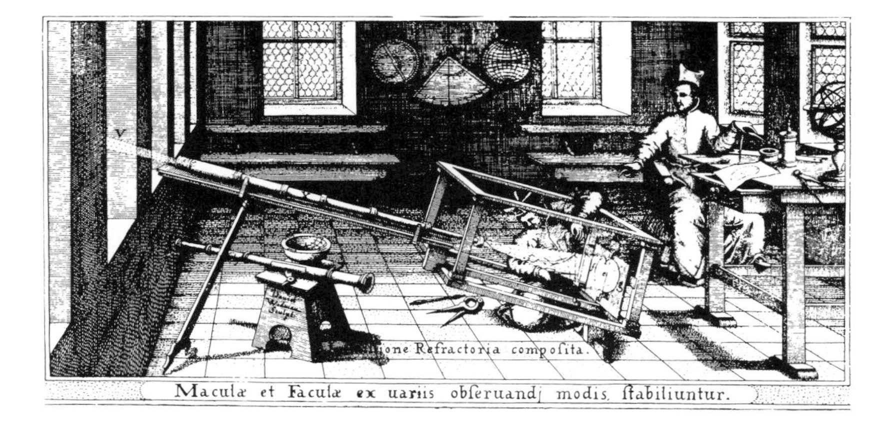

Figure 2.2. Top: Using a telescope, Scheiner projected an image of the Sun onto a white screen. Bottom: The same method was used by Hevelius (1611–1687). The drawing shows him using a projected image of the Sun to observe a solar eclipse

protection. Even before the birth of Christ, there are records of dark spots seen on the Sun. It is not always easy to make sense of some of the old accounts, such as, for example, when we read that in the year 354 BC a dark object was seen against the disk of the Sun 'as big as a hen's egg'. In old Chinese sources there are 45 reports of sunspots for the period between 301 BC and AD 1205. Einhard, the biographer of Charles the Great, wrote that in the year 807, the planet Mercury appeared as a dark spot and remained in front of the Sun for eight days. That must have been a large sunspot, because Mercury cannot stay in front of the Sun for more than about half a day. Similarly, Kepler, in Prague, saw a dark spot on the Sun on 28 May 1607, which he took to be the planet Mercury. As he confided in a letter to David Fabricius, it was only later, after he had learned about the existence of sunspots, that he realized that he had actually seen a large spot. So there is no doubt about the fact that there have been sunspots for hundreds of years.

THE MOTION OF SUNSPOTS

Very early on it was noticed that, over a period of days, sunspots slowly moved across the visible face of the Sun. A sunspot observed on one day is slightly farther west a day later. Scheiner's drawing, reproduced in Figure 2.3, shows how spots move across the disk of the Sun. Over about 13 days, a

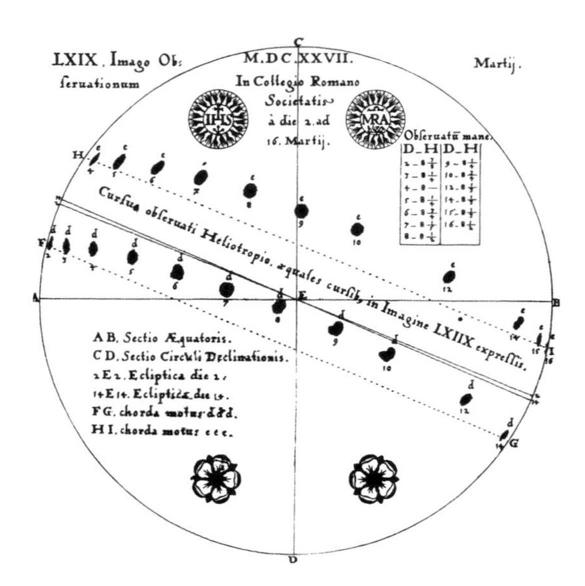

Figure 2.3. In Scheiner's drawing from the year 1627, two spots are shown as they were observed at daily intervals. Note how they move from the left (eastern) limb towards the right-hand, or western limb. They thus reflect the rotation of the Sun

spot moves from the eastern edge (or *limb*) to the western. It was also noted that as it does so, it is distorted by perspective, because at the limbs it is seen at an angle. Even if it were perfectly circular, it would naturally appear elliptical when seen obliquely. This effect may also be seen from Scheiner's drawing of 1630.

Why do the spots move across the disk? What we are seeing is the rotation of the Sun. Spots do not move at a constant speed across the disk: when they are at the limbs, they appear to move more slowly towards the west. Their motion is fastest when they are at the centre of the disk. But this is only to be expected, if we are observing a sphere that rotates evenly around its axis. At the limbs, the spots are moving more or less directly towards or away from us, and their motion is not readily visible. When a spot is in the centre of the disk, its motion is at right-angles to our line of sight, and it appears to move relatively rapidly. Sunspots therefore show us that the Sun rotates once on its axis in about 27 days. Frequently, a spot that has disappeared at the western limb, reappears in the east 13 days later. Sunspots appear and disappear, and many die only hours after they have been born, whereas other may last for weeks. A few even persist for three or four rotations.

The motions of sunspots have taught us a lot more about the rotation of the Sun. They have shown how the Sun's axis is not perpendicular to the plane of the Earth's orbit. During the course of a year we sometimes see slightly over the north pole of the Sun, and, half a year later, over the south pole (Figure 2.4).

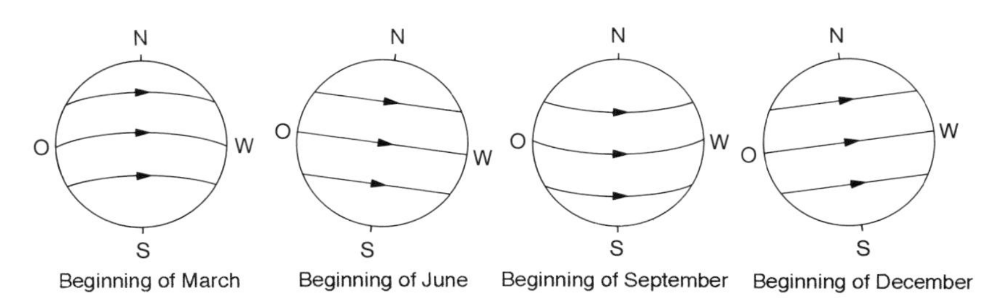

Figure 2.4. Because the Sun's axis of rotation is inclined relative to the plane of the Earth's orbit, during the course of a year we see the rotation of the Sun from slightly different directions. In March we see rather more of its southern hemisphere, and in September more of the northern

WHAT ARE SUNSPOTS?

The pictures that Scheiner published as early as 1630 show that a sunspot has a dark centre that is surrounded by an area that is not quite so dark. The centre is now known as the *umbra*, and the surrounding area, the *penumbra*. (Modern pictures, such as Figure 2.5, clearly show the umbra and penumbra.

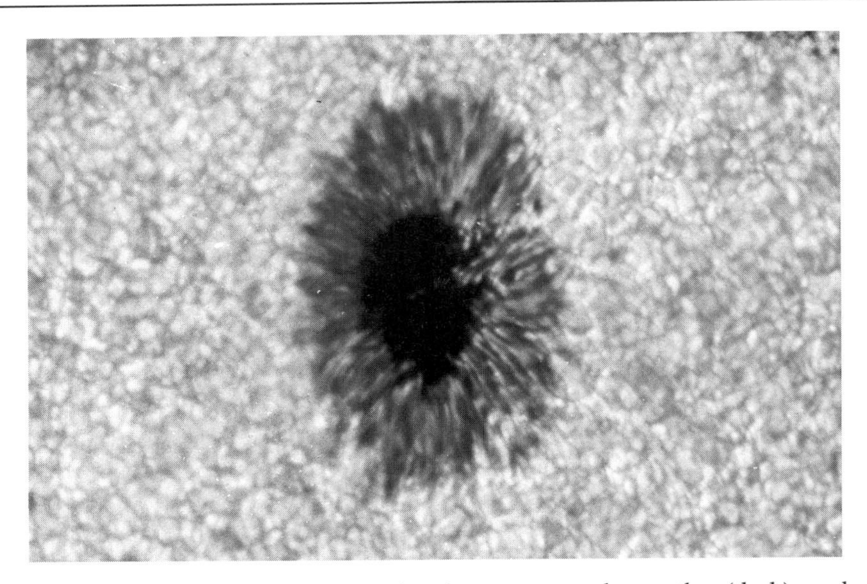

Figure 2.5. This modern photograph of a sunspot shows the (dark) umbra, surrounded by the penumbra, which consists of filaments that extend radially outwards from the centre of the spot. Around the spot, the granular structure, or granulation, of the undisturbed surface may be seen. The spot is not in the centre of the solar disk; the closest limb is outside the picture to the right. We are therefore seeing the nearly circular spot at a slight angle, so it appears elliptical. Because of the Wilson Effect, which is described in Figure 2.6, the penumbra in the left half of the spot appears narrower than that on the right. (Photo: H. Wöhl, taken with the Newton Vacuum Telescope of the Kiepenheuer Institute for Solar Physics, Freiburg, Germany, which is sited on Tenerife)

At the same time the penumbra shows a filamentary structure.) No one had previously expected the brilliant disk of the Sun to have dark spots, and no one could understand the phenomenon. Physics was then in its infancy, and even today astrophysicists have no theory that fully explains all the properties of sunspots.

In 1779, a sunspot visible to the naked eye aroused William Herschel's interest. That was two years before this famous astronomer discovered the planet Uranus. Herschel tried to explain sunspots on the assumption that the body of the Sun was actually cold. He thought that it was only the hot atmosphere of the Sun that was responsible for the Sun's radiation. He believed that sunspots were holes in the glowing layer of clouds, through which the dark ground was visible. He also tried to explain the umbra and penumbra: beneath the glowing layer of clouds there was a cooler haze layer, which, even if it did not itself glow, reflected the light from the higher, hot layer. The fact that the penumbra must lie at a lower level than the surrounding surface, and the umbra even deeper, had been indicated by a feature of sunspots that had been discovered as early as 1771 by the

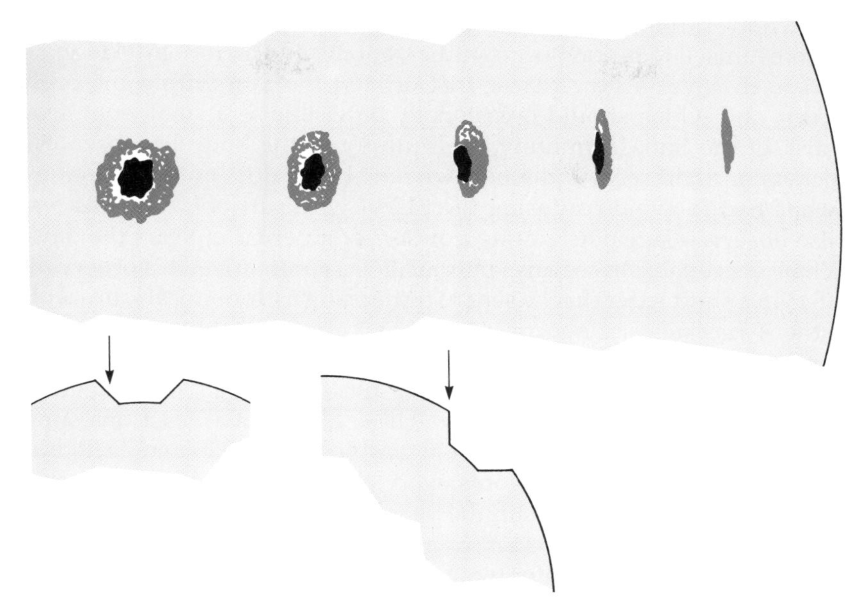

Figure 2.6. The cause of the Wilson Effect. If a spot is not at the centre of the solar disk, the portion of the penumbra closer to the centre appears narrower than the region closer to the limb. This appearance arises because we see down into deeper layers of the Sun in the umbra than in the surrounding area. This is made clearer in the two lower diagrams. On the left we observe the spot in the direction of the arrow. When the spot is close to the centre of the disk, it appears symmetrical. On the right, we see how the area of the penumbra closer to the centre of the Sun appears reduced because of perspective

Würtemberg priest Ludwig Christoph Schülen (1722–1790). He noticed that as a spot approached the limb, the penumbra changed in a way that suggested that the umbra was deeper than the glowing surface of the Sun. In other words, as if the umbra lay at the bottom of a depression, and the penumbra were the sloping walls (Figure 2.6). Three years later, and independently of Schülen, the Scots astronomer Alexander Wilson (1714–1786) discovered the same phenomenon. Nowadays it is generally known as the *Wilson Effect*. About three quarters of all sunspots show this effect as they move from the limb towards the centre of the disk, or from the centre towards the opposite limb.

THE SUNSPOT CYCLE

Although the source of the Sun's energy was not yet known, further regularities were discovered. The most important discovery was made by a pharmacist from Dessau, now in the German state of Sachsen-Anhalt.

In 1829, Heinrich Samuel Schwabe (1789–1875) sold the pharmacy that he had taken over from his grandfather, to 'begin his real life'—in other words, to devote himself to his favourite studies, botany and astronomy. He actually hoped to discover a new planet that orbited the Sun within the orbit of Mercury, and which should occasionally be visible as a small spot against the disk of the Sun. In hunting for 'Vulcan', as the body had been called (which no one had seen, and which we now know does not exist), naturally Schwabe had to avoid confusing the planet with sunspots. For this reason he also observed sunspots, noting down over a period of years the days on which he was unable to see any spots at all. He summarized his observations for 1843 in a short letter that he sent to the leading astronomical journal of the day, the *Astronomischen Nachrichten*. He had already published his data for the years 1826 to 1837 in the same journal. Now however, looking back over a total of 17 years, he realized that his results showed a striking regularity. Because there were some years in which he had seen at least one sunspot every day he had observed the Sun. Examples were 1828 and 1829, but he had also been able to see sunspots every clear day during the years 1836, 1837, 1838, and 1839, whereas in the years around 1833 and 1843, the Sun's disk had been free of sunspots for over 100 days. From this he concluded that sunspots occurred with greater frequency at intervals of about 10 years, and that during the intervening years they were less frequent.

Not much attention was paid to Schwabe's discovery at first. It was only in 1850, when Alexander von Humboldt mentioned Heinrich Schwabe's work in the third volume of his work *Kosmos*, in which he summarized the current scientific picture of the universe, that the world became aware of the amateur astronomer from Dessau. In his book, Humboldt printed Schwabe's table showing the number of spot-free days per year. Since Schwabe had published his data, another seven years had passed, and Schwabe was able to provide Humboldt with additional results for this period. The expanded list showed that Schwabe had also found no spot-free days in the years 1847, 1848, and 1849. That agreed precisely with the spot-frequency period that he had described previously. Approximately every 10 years, spots were so frequent that over a whole year there were practically no spot-free

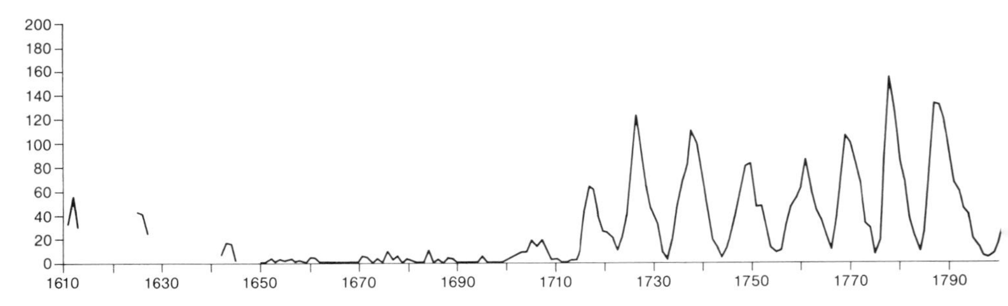

Figure 2.7. Sunspot numbers show an 11-year cycle. The latest sunspot maximum occurred in early 1990

days. Humboldt wrote 'No living astronomer, despite being equipped with excellent telescopes, has been able to devote such continuous attention to this phenomenon. Over the long period of 24 years, Schwabe has frequently examined the Sun more than 300 days in the year. Because his observations of sunspots from 1844 to 1850 have not yet been published, so I am indebted to his kindness in providing me with these, and also to answering a number of questions that I had asked.'

This was how the 61-year-old amateur astronomer first achieved a scientific reputation. The Royal Astronomical Society of London awarded him their Gold Medal. Despite the fact that, from the age of 41, he suffered from gout every winter, he was able to observe yet another two sunspot maxima. They occurred with the regularity that he had predicted.

We now know that sunspots occur most frequently with a period of about 11 years. Only later investigations showed us how regular the solar cycle is. They also showed us, however, that it does always follow this same pattern (Figure 2.7).

As a measure of sunspot activity, Schwabe used, on the one hand, the number of sunspot groups that he observed, and on the other, the number of days in the year on which he observed the Sun and did not see any spots.

When people wanted to examine this regularity in more detail, they wondered whether the number of visible sunspots was a suitable measure of solar activity, or whether it should be the number of *groups* of sunspots. What counted, spots or sunspot groups? It was a Swiss astronomer who found a compromise solution.

Rudolf Wolf (1816–1893) from Fällenden near Zürich, was originally a mathematician at Bern, but in 1847 he became director of the observatory there, and later accepted the chair of astronomy at Zürich. He established an international solar patrol, which aimed to observe the Sun for spots on every possible day of the year, and from as many places on Earth as possible. It became essential to find a measure of sunspot activity on which everyone could agree. Wolf devised the *Relative Sunspot Number*, which has now been used for over one hundred years. This number is established as follows:

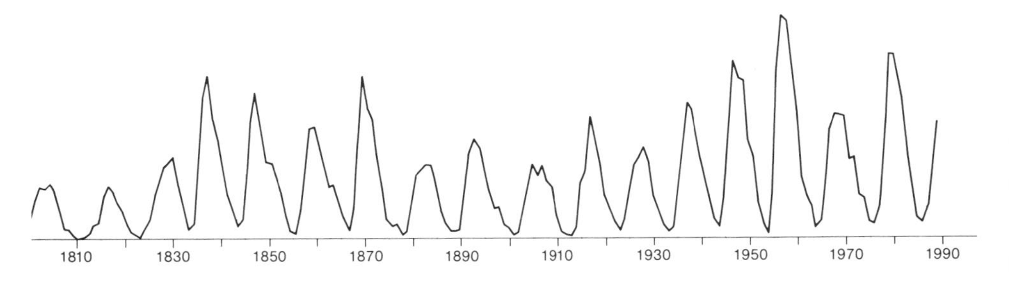

Figure 2.7. (*continued*)

All the groups of sunspots visible on the Sun are first counted, and then all the individual spots, whether they occur singly, or in one of the groups already counted. The number of spots is added to ten times the number of groups. (This is actually a very simplified description of the procedure used to establish the relative sunspot number.) If no spot is visible on the disk, then naturally the numbers of groups and individual spots are both zero, as is the relative sunspot number. The higher the number of spots and groups, the higher the relative number. On days with particularly strong solar activity, it may reach a value of 300. Wolf also succeeded in reconstructing relative numbers back to 1730, using old data.

These clearly showed the cyclic nature of the number of sunspots, as found by Schwabe. It should be noted, however, that the number may fluctuate considerably from day to day. We only have to think of what happens when a group of some 30 spots disappears over the solar limb. The relative number drops suddenly by 40, without anything special having happened on the Sun. The disappearance of a group has nothing whatsoever to do with the Sun, it simply means that it is no longer visible from Earth, because we are unable to see round behind the Sun. To avoid this random effect, the relative numbers shown in Figure 2.7 are averaged over a whole year.

THE BUTTERFLY DIAGRAM

In the middle of the last century other regularities in the behaviour of sunspots were discovered. We have an English amateur astronomer, Richard Christopher Carrington, who observed the Sun from his private observatory, to thank for various important discoveries. He was the son of a rich brewer, and originally meant to study theology, but became more interested in astronomy. After a period of three years learning to be an observer, he built his own observatory. Schwabe's discovery had just become known, and Wolf had been able to reconstruct the relative sunspot numbers back into the eighteenth century. Sunspots were ideal objects for our novice astronomer to study. He was to be extremely successful.

From the motion of spots across the disk of the Sun, he first discovered that the Sun does not rotate as a solid body. For instance, while a spot close to the equator takes less than 25–26 days to complete one rotation, a spot at latitude 30° takes about 27 days. At higher latitudes, for example at 80° north (or south), a point on the surface of the Sun takes more than 30 days to complete one rotation (see Figure 2.8). Because sunspots, as we shall shortly see, only occur close to the equator, the rotation of polar zones of the solar surface has been determined by other methods (see Chapter 5).

Whereas the discovery and confirmation of the eleven-year period in the relative sunspot numbers depended on *counting* sunspots, Carrington determined the law governing the rotation of the Sun by *measuring* the positions of individual spots.

In doing so, he discovered another regularity. The spots prefer different zones during the course of a cycle. At the time of sunspot maximum, they are generally found in two bands running parallel to the equator at about latitudes 15° north and south. As their number declines in subsequent years, they primarily occur at lower latitudes. When sunspot minimum is nearing its end, the new cycle's spots appear in two bands that are significantly farther away from the equator. Their latitudes are about 35° north and south. There may still be a few spots close to the equator, but they continue to decline in number. Instead, more and more spots appear in the two high-latitude zones. The positions at which spots occur at the beginning of a cycle, i.e., just

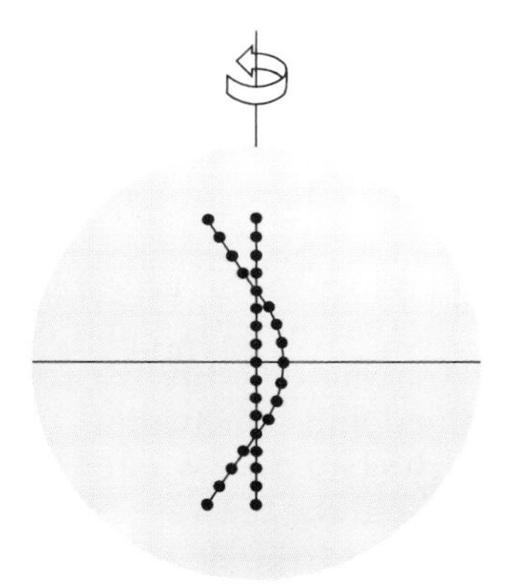

Figure 2.8. The Sun does not rotate as a solid body: points close to the equator move faster than those at higher latitudes. If we were able to mark its surface along a meridian (shown here as dots along a straight line through the centre of the diagram), after one rotation the points on the surface would appear approximately as shown by the curved line

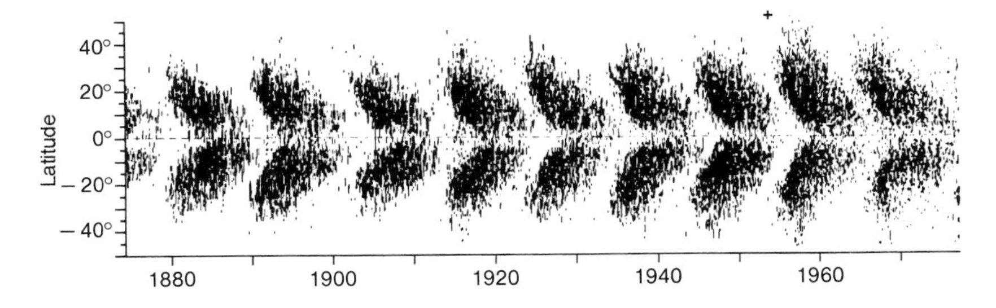

Figure 2.9. The 'Butterfly Diagram' shows the heliographic latitude (i.e., the distance from the equator in degrees) of the positions of sunspots. Following sunspot minimum, spots appear in the middle latitudes in both hemispheres, around 30° North and South. Over the following years, the spots occur closer to the equator, and their numbers eventually decline. At about the same time new spots occur at higher latitudes, beginning two new 'butterfly wings'. The position of the short-lived region in 1953, which is mentioned on p. 89, is marked by a cross

after minimum, are much farther from the equator than they are towards the end. During the course of a cycle, the zones in which spots occur slowly migrate towards the equator. When the cycle is approaching minimum, the two zones are close to the equator. This may be seen in Figure 2.9. Because of its appearance, this sort of plot is known as a *Butterfly Diagram*. It allows us to distinguish between spots belonging to the old cycle and those belonging to the new.

THE TRAGEDY OF RICHARD CARRINGTON

Before I go any further, I would like to linger a little longer on Carrington. He was one of the most important solar physicists of the nineteenth century, and did not just discover the Sun's differential rotation, and the butterfly diagram. He was the first to see a flare on the Sun (see p. 97), and he was also the first to suspect that magnetic-field events on Earth were related to activity on the Sun. (We shall return to this aspect in Chapter 11.)

Even great scientists are human, however, and often suffer cruelly at the hands of fate. The end of Carrington's life was overshadowed by events that have never been explained, but which allow us to glimpse the tragedy suffered by this great man, who had once wanted to be a priest.

In 1865 he became very ill, sold the brewery that he had inherited, and retired to his house in Surrey. There, he began to build a new observatory. At about this time, someone attempted to murder his wife, Rose Ellen Carrington. The perpetrator was apprehended, and condemned to 20 years imprisonment.

On 17 November 1875 Mrs Carrington was found dead in her bed. She had apparently died from an overdose of her medicine. At the inquest that was held shortly afterwards it was established that Carrington, who had, as usual, given her the medicine the previous evening, had seen his wife lying on her face when he woke the next morning, but had thought that she was asleep. It was only later that the maid noticed that she was dead, although her body was still warm. The family doctor, who confirmed the death, stated that the dose of medicine administered by Carrington was not dangerous. Analysis of the stomach contents showed no signs of poison, and so the cause of death was not established. Despite this, the coroner reprimanded Carrington for lack of care for the invalid. This indirect verdict of guilty must have hit him hard. On the same day he left his house, and only came back a week later. In the meantime, the servants had left. Carrington returned to the house on 27 November. No one saw him alive after that. When anxious neighbours broke down the doors, they found his body, lying on a mattress in a locked room. The cause of death was stated to be a stroke, but there was a rumour that it was suicide.

THE HIGH-SCHOOL TEACHER WHO BECAME A SOLAR PHYSICIST

Amateurs have played a major role in the study of sunspots. There was the doctor Johannes Fabricius, the pharmacist Schwabe, the mathematics teacher Rudolf Wolf, and the privately tutored Carrington, originally intended to be a theologian, who also had to become involved in the brewery that he inherited. While Carrington was working in his personal observatory, the high-school teacher Gustav Spörer (1822–1895) was investigating sunspots from his home at Anklam near the German Baltic coast. He knew nothing about Carrington's measurements of solar rotation, and discovered the Sun's remarkable rotation independently. Later, he was also able to confirm the butterfly diagram from his own observations. In 1874, this teacher from Anklam was appointed observer at Potsdam, where he worked until shortly before his death. His discoveries acted as a spur to the establishment of the Astrophysical Observatory at Potsdam in 1879. It was here that he recognized one of the most remarkable irregularities in the solar cycle. We shall return to this particular point later.

With the foundation of the Potsdam Institute, professional astronomers became involved with the Sun, and they have since investigated it with spectrographs, radio telescopes and space probes—not that amateurs have abandoned it. Sunspots are in fact very profitable objects for observers that have only small instruments. This is shown, for example, by the butterfly diagram that Frau Sieglinde Hammerschmidt, a housewife from Solms near Wetzlar, obtained from her own observations. To gain some idea of her achievement, it must be remembered that it is not enough to simply project the disk of the Sun and draw every spot on a sheet of paper. To plot a spot or a group on the diagram, it is also essential to take into account the fact that during the year we see the Sun from slightly different angles (Figure 2.4). Figure 2.10 shows Frau Hammerschmidt's butterfly, obtained with observations on a total of 1224 days between 17 January 1976 and 24 December 1986, and which includes 6701 individual observations. It will be seen that the last cycle ended in 1986, and that in the second half of that year the first spots of the new cycle appeared at high latitudes.

The individual cycles have been numbered, beginning with number 1, which is the cycle that reached maximum in 1760. Cycle 21 therefore ended in 1986, and Cycle 22 began. The latter was reaching maximum around the time I finished work on the original German text of this book.

After the discovery of photography, astronomers had the chance of obtaining pictorial records of sunspots. On 2 April 1845, at the Paris Observatory, the French physicists Armand Hippolyte Louis Fizeau (1819–1896) and Léon Foucault (1819–1868) managed to obtain a Daguerreotype of the disk of the Sun. From then onwards it was possible to photograph sunspots

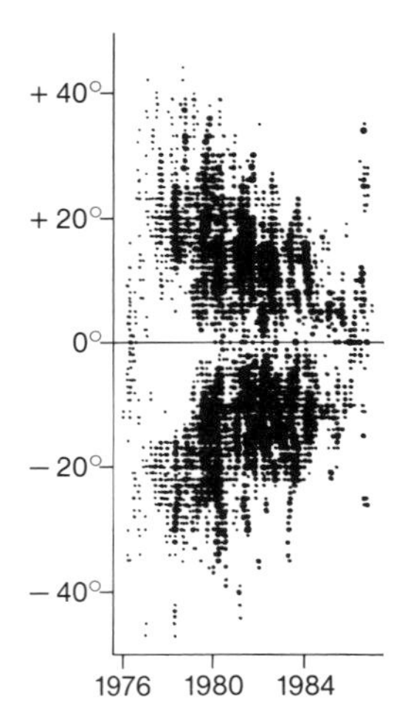

Figure 2.10. This diagram shows the 'butterfly' for Cycle 21, which followed the last cycle included in Figure 2.9. It is the result of 11 years of work by Sieglinde Hammerschmidt, who is an amateur astronomer

on a regular basis. Their positions could be measured from the images, and photographic records of the constantly changing spots could be assembled.

THE SUN KING AND SUNSPOTS

Immediately after sunspots were discovered, they were eagerly observed. Within a short space of time, however, the excitement died down. This was not because sunspots had lost their initial appeal; the Sun itself was to blame. Despite the fact that sunspots may be observed with relatively simple methods, it is not really surprising that the next important facts about them were only discovered in the last century, because for about 70 years sunspots were missing.

When, at the end of the last century, Rudolf Wolf tried to reconstruct relative sunspot numbers from old observational data, he went back to the year 1700. He never commented about why he did not compile relative sunspot numbers for earlier years. He probably felt that the sources were not sufficiently reliable. The American solar researcher John A. Eddy has a different explanation: Wolf, motivated by Schwabe's discovery of the sunspot cycle, did not simply want to follow the future regular increase and decrease in the number of sunspots and sunspot groups as accurately as possible. He also wanted to prove that Schwabe's cycle could be traced far back into the past. This he was able to do back to about the beginning of the eighteenth century. Wolf probably then realized that earlier observations no longer agreed with Schwabe's cycle. He therefore mistrusted the few sources that he was able to find that dated back to the seventeenth century, and so he decided not to pursue his investigations any further back in time.

There are major reasons for believing that there were practically no sunspots at that time. This is where Gustav Spörer returns to our story. In 1889, he pointed out that the normal sunspot cycle had been interrupted for a period, which ended around 1716. A year later, the English solar researcher Edward Walter Maunder (1851–1928), who was studying the Sun at the Royal Greenwich Observatory, confirmed Spörer's suspicion. His article 'An extended sunspot minimum' appeared in 1890. Since then, that period of low sunspot numbers has been known as the *Maunder Minimum*. Eddy tried to reconstruct the relative sunspot numbers back beyond 1700. He obtained the values shown on the left-hand side of Figure 2.7.

In fact, the Sun appears to have shown hardly any spots during the reign of Louis XIV, the Sun King, i.e., between 1638 and 1715. It is certainly difficult to find material about the current solar activity for years in which no one systematically monitored the Sun—and during that period sunspots are hardly mentioned, although telescopes were available. It was possible to buy them then, and it does not require a very refined form of telescope to see sunspots. During the time in which practically no sunspots were reported, a series of other discoveries were made with the telescope. People realized that there was a division in Saturn's ring, five satellites of Saturn were discovered, and Mercury and Venus were seen as black dots that passed across the disk of the Sun. Despite this, there are hardly any reports of sunspots observed in the period between 1645 and 1715. Within this period there is even a span of 32 years during which there is not one single reported sunspot.

When Giovanni Domenico Cassini (1625–1712), the Director of the Paris Observatory, actually discovered a sunspot in the middle of the Maunder Minimum, it prompted a long note, followed by a extensive report on the last sunspot, 11 years earlier. After all, anyone who had not seen a sunspot needed to be told what one looked like. Picard, whom we have already met in discussing the measurement of the Earth (p. 2) also discovered a sunspot at about that time, and Cassini wrote that Picard was delighted, because for 10 years he had not seen any spots at all.

Other indications point to an extended sunspot minimum. We shall see that in years when solar activity is high, the number of aurorae is also high. They are produced by material that is ejected from the Sun at a particularly high rate around the time of sunspot maximum. The period of reduced sunspot frequency is also reflected in historical reports of the aurora. The strongest evidence for irregularities in the past level of sunspot activity comes, however, from a completely different source.

SUNSPOTS AND RADIOACTIVE CARBON

The Earth is bombarded by a continuous stream of charged particles coming from the depths of space. These *cosmic rays* were discovered in 1915, high in the Earth's atmosphere, by the Austrian physicist Viktor Franz Hess (1833–1964). Particles of this radiation turn atmospheric nitrogen into a carbon isotope, ^{14}C (carbon-14). The atoms of this type of carbon differ from those of normal carbon, ^{12}C, in being slightly heavier, because their nuclei contain two extra neutrons. In addition, ^{14}C is radioactive. For any given initial number of ^{14}C nuclei, half decay into oxygen nuclei every 5730 years. If cosmic rays were suddenly to cease, more and more atoms of ^{14}C would decay, until eventually none were left. But if cosmic rays remained at a steady level for millions of years, then a certain, specific number of atoms of radioactive carbon would be formed, and this number would be such that the number of nuclei that decayed every second would equal the number of new nuclei produced. If the level of cosmic rays varied with time, then the numbers of ^{14}C atoms would also vary.

The radioactive carbon atoms mingle with ordinary carbon and, because they do not differ chemically, they are taken up, in the form of carbon dioxide, by plants, and stored within them (in tree rings, for example). If the individual annual rings in an tree are investigated, it is thus possible to determine the ratio between normal and radioactive carbon for any given year. But what has this to do with sunspots?

When the Sun is very active, the material that continuously streams away from its surface is accompanied by magnetic fields, which deflect cosmic rays from the Earth's vicinity, preventing them from reaching the atmosphere. So when solar activity is at a maximum, less ^{14}C is produced in the Earth's atmosphere. Any tree rings laid down at that time contain less radioactive carbon. Tree-ring analysis allows solar activity to be traced far back into the past. The results are shown in Figure 2.11. Two periods of high ^{14}C levels may

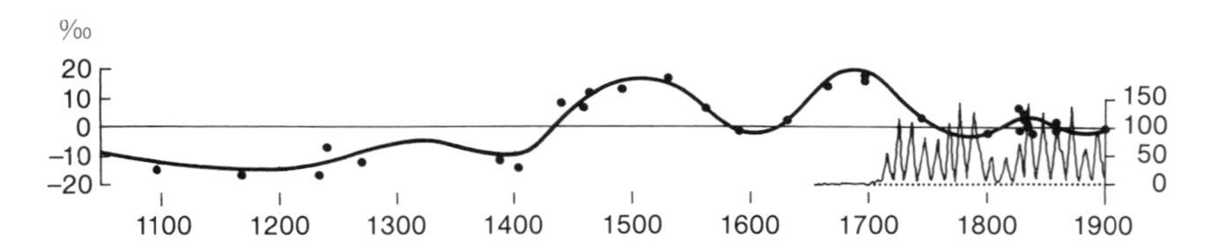

Figure 2.11. The past level of radioactive ^{14}C as shown by the analysis of the wood from tree rings. From the beginning of the eighteenth century, relative sunspot numbers are also shown, with the appropriate scale on the right. The heavy line shows the variation, in parts per thousand (per mill) of total carbon, in the number of radioactive ^{14}C atoms in a sample. It shows that the relative abundance of ^{14}C was extremely high during the Spörer Minimum around 1500 and during the Maunder Minimum around 1700 (after J.A. Eddy)

be clearly seen: the Maunder Minimum in the second half of the seventeenth century, and a second one, which has been called the *Spörer Minimum*. That appears to have occurred between about 1460 and 1540. There are, moreover, practically no reports of aurorae during this period.

Using the ^{14}C method, the strength of cosmic-ray activity may be traced back from the present to about 7000 years ago. At the very first glance it is obvious that, thousands of years ago, there was more radioactive carbon in the air than at present. This is related to the Earth's magnetic field. We now know that the Earth's magnetic field has reversed its direction many times. The weak magnetic fields emanating from the Sun may control the stream of cosmic rays reaching the Earth, but the Earth's magnetic field itself is even more important.

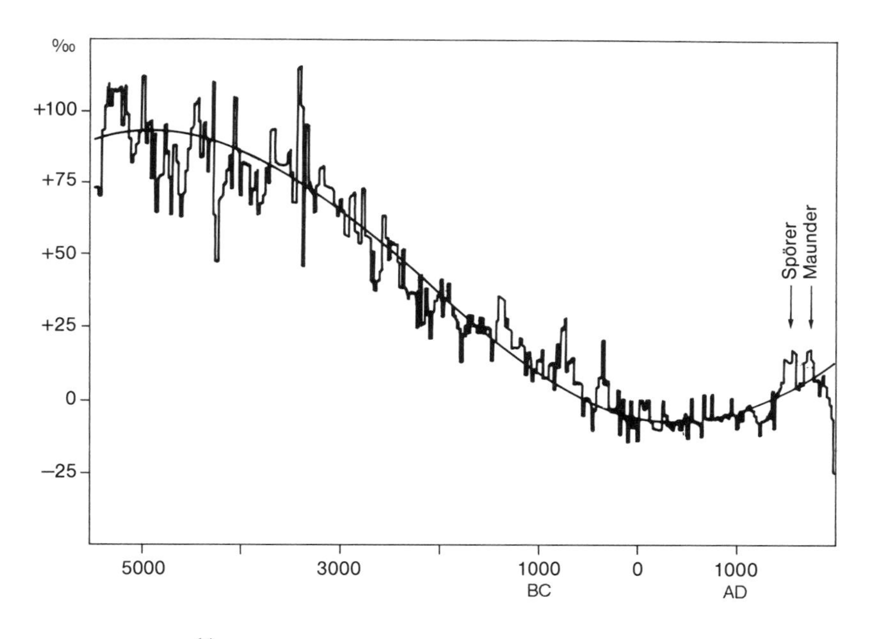

Figure 2.12. The ^{14}C abundance (plotted as in Figure 2.11) has shown very strong variations over the last 7000 years, and these variations are linked with the Earth's magnetic field. Seven thousand years ago, the Earth's field was significantly weaker, so the Earth had less protection from cosmic-ray particles, and the ^{14}C production rate was high. When the Earth's field is strong, as during the last 3000 years, many cosmic-ray particles are deflected, and the ^{14}C content is low. Magnetic fields from the active Sun also help to shield the Earth from cosmic-ray particles and thus lower the atmosphere's ^{14}C content. During the Spörer and Maunder Minima the magnetic fields originating in the Sun were missing, and the ^{14}C content increased significantly. The decline at the right-hand side of the diagram (which actually corresponds to the year 1950), has been caused by the carbon released into the atmosphere by human activity. This has added ordinary carbon to the atmosphere and 'diluted' the relative abundance of the ^{14}C isotope that is produced by cosmic rays (after J.A. Eddy)

Note, in Figure 2.12, the decline in ^{14}C abundance since the beginning of the twentieth century. This reflects the burning of fossil fuels. Carbon, which was absorbed from the air by plants long ago, is being released back into the atmosphere. The radioactive ^{14}C has decayed long ago. So our chimneys are pouring non-radioactive carbon back into the atmosphere. The ^{14}C created by current cosmic-ray levels is being diluted by the non-radioactive fossil carbon.

Heinrich Samuel Schwabe discovered the 11 year sunspot cycle. It was only in this century that it was found that the cycle was really 22 years long. This arose from more detailed studies of sunlight.

3

SUNLIGHT

> Goethe had borrowed a few prisms from a teacher at Jena and intended to experiment with them. They lay, forgotten, in his cupboard. The teacher sent him a reminder and, finally, a messenger. Goethe handed over the pieces of polished glass without hesitation, but at the last moment, literally in passing, he took up a prism . . . and briefly held it against the wall. . . . And saw that there was no display of brilliant colours! He could see only white against the white wall. Enlightenment hit him like a flash: Newton's theory was wrong.
>
> Richard Friendenthal, *Goethe*

In the flickering light of the torch, it seemed as if the animals on the rock walls were moving. They were oxen, drawn with a few lines of black and red, without any naturalistic detail. The nameless artist had made use of irregularities in the cave wall, turning a bulge into the belly of an animal. I was suddenly conscious of how an unknown artist was capable of affecting my emotions, more than 20 000 years later, and how I was feeling what he intended.

My inner emotion was caused by what I saw. We look at pictures in museums, enjoy the sight of sunset over the sea, and love the view over a beautiful landscape: all because we have the gift of sight.

Light falls on our eyes, is recorded in the nerve cells of the retina, and carried by the optic nerve to the computer inside our brain. There a picture is formed that speaks to our soul. Light brings us these influences, but what is it?

We know that it falls on our eyes from outside. But this was not always the view. Plato (427–347 BC) thought, for example, that rays of vision went out in straight lines from our eyes, and that we experienced something when they encountered an object. Even today we still use the expression 'cast one's eyes on'. A young man who 'casts his eye on a woman' is thus Platonic in a double sense: as far as seeing is concerned and (of course!), in every other respect as well. It is difficult for us to appreciate Plato's ideas on this—however great he may have been in other fields.

THE SPECTRUM

Around 1670, Newton carried out his crucial experiment with a prism, that is, with a triangular piece of glass, like that shown in Figure 3.1. He describes this in his book *Opticks*, which appeared in 1704: 'In the Sun's Light let into my darken'd Chamber through a small round Hole in my Window-shut, at about ten or twelve Feet from the Window, I placed a Lens, by which the Image of the Hole might be distinctly cast upon a Sheet of white Paper, placed at the distance of six, eight, ten or twelve Feet from the Lens. For, according to the difference of the Lenses I used various distances. . . . Then immediately after the Lens I placed a Prism, by which the trajected Light might be refracted either upwards or side-ways. . . .' Instead of a point of light, Newton saw an elongated strip. This consisted of innumerable, circular, mutually overlapping images of the hole, which all had different colours. At the top, the strip was violet, and at the bottom red. This became even clearer when he covered the hole in the shutter with a piece of paper in which he had cut a slit, which was parallel to one of the edges of the prism. Now the different coloured images of the slit lay alongside one another, and mutually overlapped. The prism had split white sunlight into all the colours of the rainbow.

Newton's bright strip of light, which consisted of a succession of superimposed images of the tiny slit, is known as a *spectrum*. Since that time, astronomers have learned how to use it to determine, not just the temperature of the luminous surface of the Sun, but also its velocity, its chemical properties, and even the strength and direction of magnetic fields that are invisible to our eyes. Naturally, Newton had no idea of all this.

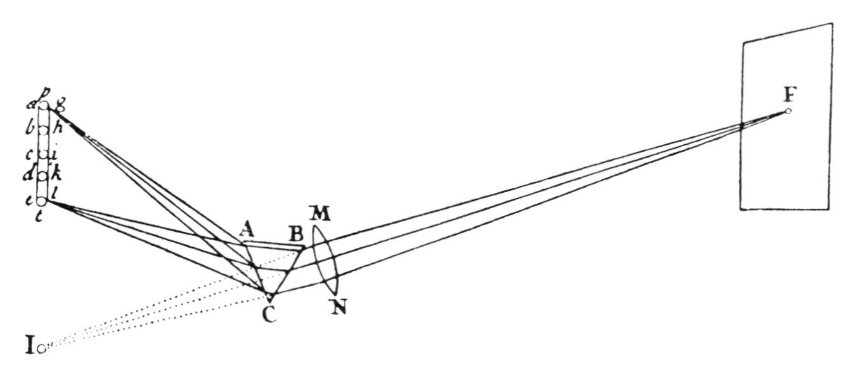

Figure 3.1. Newton's drawing of his experiment with a prism, a lens, and a hole in a shutter. A ray of sunlight passes through the hole F and falls on the collecting lens (MN), which would form an image of the hole at I, if there were no prism ABC, which deflects the light, violet being more affected than red light. This creates an image of the hole in the shutter for each of the different colours (top left), with red at the bottom and violet at the top. The images combine to create a coloured band of light—a spectrum

To learn more about the nature of sunlight, he took a second prism, which he placed in the beam of light, facing in the opposite direction to the first prism—i.e., one with its apex upwards and the other downwards. The second prism recombined the individual coloured rays of light. The light falling on the paper reverted to being white.

From this experiment, in which he had split white light into various colours, which he could recombine into white light, he concluded that white light consisted of different coloured components. The ideas that Newton developed about the nature of light enabled him to explain the colours of the rainbow. Within tiny raindrops, light is reflected from the rear surface. In doing so, the light must pass twice at an angle through the surface of the liquid, which therefore acts like a prism.

But what is light? Newton believed that it consisted of innumerable, tiny, coloured particles (sometimes called 'corpuscles') that were emitted at high speeds from a source of light, such as the Sun. Taken together, they appear white to our eyes. A prism is, however, able to separate them according to their colour. It deflects the violet corpuscles from their original paths more strongly than the red ones. When they pass through a second prism, with its apex in the opposite direction, and are brought back together, they appear white to us.

This view of light as corpuscles does not, however, explain all the properties of light. But before we come to this, let us consider the light that Newton did not see.

INVISIBLE LIGHT

The spectrum that Newton had obtained using a slit in his shutters, a prism and a lens, contained more than he suspected. This was proved by an English astronomer, who originally came from Hannover in Germany.

William Herschel, who was already world-famous for his discovery of the planet Uranus, frequently observed the Sun with his telescope, to the eyepiece of which he had added a coloured filter, which should have protected his eyes from the brilliant solar radiation. He realized, however, that even with filters that did not transmit any light, he could still frequently feel a sensation of warmth in his eye. He therefore suspected that the Sun's heat did not reach us in the form of visible light, but in some form that was invisible to the eye. This idea led him to conduct an experiment that closely resembled Newton's. He allowed sunlight to enter a darkened room and, after passing through a prism, to fall on a strip of paper (Figure 3.2). At the red end of the spectrum, but outside the region in which the light, dispersed by the prism was visible, he placed three thermometers. There, in the region where no light is visible to our eyes, the thermometers showed higher temperatures. Herschel had discovered rays from the Sun that lay beyond the red light in the spectrum—*infrared light*.

Figure 3.2. William Herschel's experiment, in which he discovered the Sun's infrared radiation, is an extension of Newton's experiment shown in Figure 3.1. Instead of passing through a round hole, the sunlight enters through a horizontal opening and falls on a glass prism—the lens used by Newton is omitted. The spectrum is a broad band of colour on the horizontal surface of a table, running from the violet (left) to the red (right). Outside the visible spectrum, beyond the red end, thermometers indicate that the Sun also emits invisible, heat radiation. (Bayerische Staatsbibliotek, Munich)

Stimulated by this discovery, the German physicist Johann Wilhelm Ritter (1776–1810) exposed silver chloride to various regions of the solar spectrum. This silver compound is altered by light, which is why it, and silver bromide, are used in photography. Ritter found that the strongest chemical reaction occurred beyond the violet end of the spectrum. He had discovered the Sun's *ultraviolet radiation*.

Herschel and Ritter had discovered solar radiation that was invisible to the eye, and which extended beyond both the red and violet ends of Newton's spectrum. We now know that the spectrum extends much farther at both ends. Beyond infrared radiation there are radio waves. At the other end of the spectrum, beyond the ultraviolet, there are X-rays, and finally, gamma-rays. The Sun emits all these forms of radiation into space, but the majority do not reach the Earth's surface. Apart from visible light, the atmosphere allows radio waves through to us, even though we cannot see them. We can investigate the other radiation from space, whether from the Sun or from other celestial objects, only from aircraft or balloons, or by instruments that are launched outside the Earth's atmosphere with rockets.

LIGHT AS WAVES

According to Newton, a ray of light consisted of tiny particles that flew through space and produced the sensation of light within the eye. His contemporary, the Dutchman Christiaan Huygens (1629–1695) believed, on the other hand, that light was the oscillations in an all-pervasive medium, rather like waves on the surface of water. We now know that in fact both were right. Light behaves both as if it consisted of particles, and also as if it were a form of wave.

Because our view of the world is based on the events that occur in everyday life, we find this hard to accept. What is light really, is it waves or particles? It is neither one nor the other. If we think of a wave, we tend to think of the ripples that propagate outwards on the surface of a pond into which we have just tossed a stone. A 'particle' tends to suggest a small, round, and possibly hard sphere, rather like a grain of seed. Waves and particles appear to be incompatible. One seems to be rhythmic motion of a fluid—or, in the case of a sound wave, of a gas—and the other something material, that you could bite between your teeth. So what does it mean to say that Newton's corpuscular theory of light is just as valid as Huygens' wave picture?

The solution lies in the fact that light simply cannot be described in any clear-cut manner. It is actually a physical concept, which possesses various complicated properties that can only be described formally, using mathematics. For many of these properties, however, we can manage with a simpler mental image. But we cannot do this for all of its properties.

We should not allow ourselves to be confused because light sometimes behaves like a wave, and sometimes like a handful of seed. It seems natural to describe other things in different ways, according to which of their properties we are concerned with at the time.

To give an example: A country is like a building resting on many columns. One of these might represent its scientific reserves, and another the strength of its army. It is not such a bad image. If the country goes bankrupt, and one column collapses, it does actually behave like a crumbling building that ends up as a pile of rubble. If we speak of its relationships with neighbouring countries, however, we tend to ascribe human properties to it. We speak of friendly cooperation, mistrust, revenge, or enmity. Is a country a building, or an object with human attributes? It depends on what we are discussing, which comparison we can use. The same applies to light. Some phenomena may be explained with the wave-like picture, and others with particles.

If I ask how a beam of light travels through empty space, generally the image of particles is adequate. Imagine a wall with an opening in its centre.

If light shines on it, it intercepts the stream of particles that are moving in a straight line, and throws a shadow. The light that bounces back from the wall also agrees with the image of individual particles. However, the opening would allow both light and individual particles to pass through in a straight line. The particle image appears to explain everything perfectly adequately.

But if the opening in the screen is made extremely small, say perhaps one thousandth of a millimetre, then the light behind it no longer travels in a straight line. The particles passing through the opening appear to deviate from a straight path and reach areas behind the screen that ought to be in shadow. This is where the particle image breaks down. We shall see shortly that this, and a similar experiment, may be used to detect the wave-like nature of light, which may be explained with the wave model. Has Huygens beaten Newton? First let us consider the wave model.

WHERE NEWTON'S CORPUSCULAR THEORY FAILS

It was the English doctor and physicist Thomas Young (1773–1829) who first proved that light is a wave. Even as a student, he was fascinated by the question of how the lens in the human eye altered its shape when it looked at objects at different distances. In 1801, he discovered the astigmatism of the eye. By then, he had already been practising as a doctor in London for two years. Also around that time he began his experiments that were eventually to reveal the nature of light. We will describe the most important of his experiments.

Before doing so, however, we will mention another aspect of this versatile man. He was also interested in Egyptology. At that time, hieroglyphics had not yet been deciphered, although the British Museum contained the famous Rosetta Stone. Found northwest of Rosetta on the Nile by Napoleon's expedition to Egypt, this stone, which is about as big as a table top, is inscribed with three texts in different scripts. One is Greek, and was thus legible; another was in hieroglyphics, and the third in a form of cursive or written script developed from hieroglyphics. It was reasonable to assume that the contents of all three texts were the same. It seemed to be the ideal opportunity to finally decipher hieroglyphics. Eventually, the French historian Jean-François Champollion (1790–1832) succeeded in doing so. But the London doctor was hard on his heels, and his name is still mentioned in all the books that describe the deciphering of hieroglyphics.

Here, however, we are concerned with his contribution to deciphering light. The principle behind his crucial experiment is shown in Figure 3.3. Light from a source falls on a screen in which a tiny hole has been made. It passes through the hole and, if the latter is small enough, it does not simply continue in its original straight line, but spreads out into a fan. This seems unnatural, because we are used to seeing the light that comes through a window cast

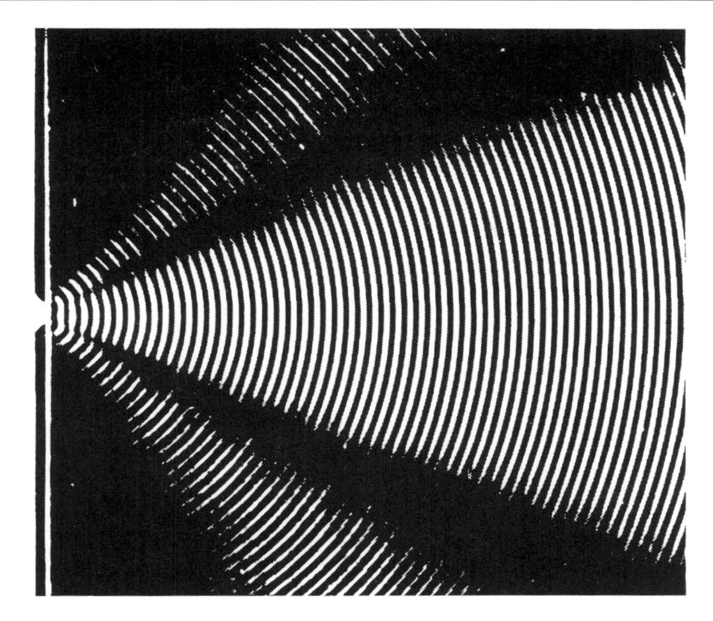

Figure 3.3. How the wave-like nature of light may be demonstrated. Waves, which are emitted at a considerable distance off to the left of the picture, pass through a tiny hole in a screen. Beyond the screen they do not propagate in a straight line, but spread out into a fan. Light always behaves in the same way, when the hole is small enough

sharp-edged shadows of the glazing bars on the floor, which is a sign that light travels in straight lines. But this is not the whole truth. Extended light-sources, such as a car's headlight, or the Sun itself, cast shadows with fuzzy edges. The sharpest shadows are produced by point sources. But no shadow is absolutely sharp. Some light is always bent into the shadow area. This cannot be explained with the particle model of light. Once a seed has passed through a hole, it continues to move in a straight line. The same would apply to Newton's corpuscles. It is different with a wave on the surface of water, however, that encounters a wall containing an opening. Behind the wall the wave spreads out in the shape of a fan.

The actual proof that light behaves like a wave was obtained by Young in another experiment. He bored two tiny, closely spaced holes in a screen. Let us first imagine this experiment carried out with water waves. Waves pass through each of the two holes and fan out on the 'shadow' side. There the water is set into motion by the two systems of waves, and a regular pattern forms, where the water is stationary in certain areas, but continuously moves up and down in others (Figure 3.4). This may be simply explained.

The situation is shown schematically in Figure 3.5, which represents an instantaneous image of the water surface. A wave approaches from the left. Wave crests are white, troughs are black. At the screen, the wave crests and

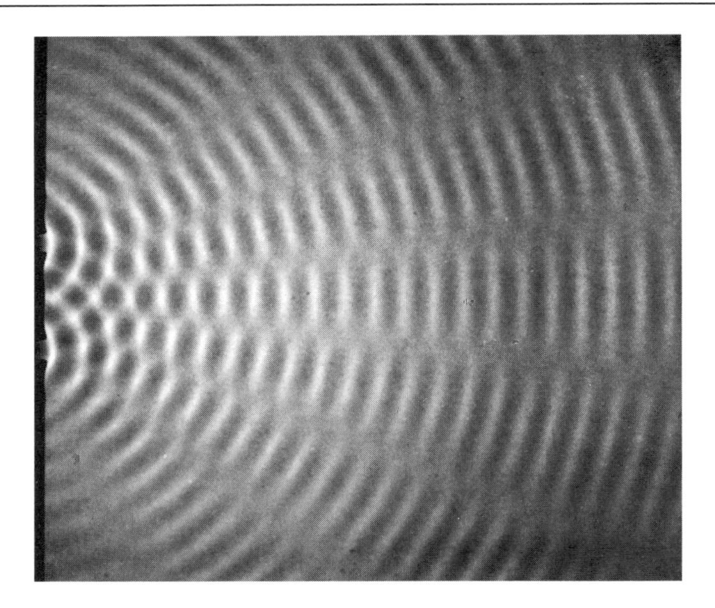

Figure 3.4. Parallel, regular waves (coming from the left of the diagram) pass through two tiny, closely spaced openings, beyond which they spread out and combine to form a regular pattern. Bands appear where the two sets of waves cancel one another out; the water remains stationary. At other points the two waves mutually reinforce one another. Thomas Young was able to prove that this typical property of waves was also shown by light (as described in Figure 3.5)

troughs follow one another in a regular rhythm. On the farther side of the screen, a fan of waves spreads out towards the right from each opening. The crests and troughs pass equally through both openings. When a crest passes through one opening, another passes simultaneously through the second. Every point of the surface of the water behind the screen will be affected by the waves passing through the two openings.

Let us first concentrate on a point that is equidistant from both openings, such as that marked A. Because wave crests pass through both openings at the same time, and the point A is the same distance from both openings, both crests also arrive simultaneously at A. The same applies to the troughs. The water at point A will therefore be moved upwards (in the crests) and downwards (in the troughs) simultaneously by the waves from both openings. The surface of the water will oscillate up and down, and both wave systems will mutually reinforce one another, because both are acting in the *same* direction. It may be very different at a neighbouring point B, which is shown slightly farther towards the top of Figure 3.5 than point A. A wave crest from the lower opening has to travel a greater distance than one from the other opening. At B the crests no longer coincide. It may be, for example, that

Figure 3.5. Waves that pass through two closely spaced openings in a screen produce two trains of waves. At point A, either crests or troughs arrive simultaneously, and both wave systems reinforce one another. At point B, a crest in one wave train arrives simultaneously with a trough from the second wave system. The two wave trains cancel each other out at point B

a crest from the upper opening always coincides with a trough from the lower. One wave is trying to raise the water surface, while the other is attempting to lower it. The surface remains stationary, because the two waves are acting in opposite directions. The regions in which the action of the two wave systems counteract one another appear in Figure 3.5 as approximately straight dark bands: where crests (white) oppose troughs (black), this diagram shows the result as black. The true appearance is actually shown in Figure 3.4, which is a picture of the effects of water waves passing through two openings.

Young demonstrated that light exhibits similar features when he repeated his experiment, allowing light from a short region of the spectrum to pass through two tiny, closely spaced holes in a screen. On the shadowed side of the screen, where the light spread out in two fan-shaped beams from the openings, he found bright and dark regions. These were the points where the waves from both openings either reinforced one another or cancelled out. This was unequivocal evidence that light was a wave!

Young's result was not readily acknowledged in England, because it contradicted the great Sir Isaac Newton, who had by then become a British institution. It was only when French scholars obtained similar results that Young's wave theory of light was accepted.

ADD LIGHT TO LIGHT AND YOU HAVE DARKNESS

Before we discuss the physicists' current view of light, we should consider this experiment that demonstrated the wave-like nature of light. It is perfectly possible for two waves to encounter and affect one another at specific points in space. This could not happen if light were simply a stream of particles, like some form of tiny spheres. Where two rays encountered one another, the density of particles would become greater. Two rays of light would only be able to reinforce one another, they could never mutually weaken one another. But Young's experiment shows that two light-waves, each of which individually appears like light, are able decrease, or even cancel one another out.

I instinctively think of the 'night-and-day light' invented by Christian Morgenstern's imagination:

> Korf found a night-and-day light,
> which, as soon as it was switched on,
> turned even the brightest day into night.

Such a light could only work if it produced light-waves that precisely cancelled out the waves of sunlight at every point in space. Because light is wave-like in nature, such a lamp is feasible, at least in principle, but it is not with the particle model.

Young proved that light is a wave, even if it was not clear what medium was altering in a wave-like manner. This wave picture was increasingly supported in subsequent years. Because the shape of the curves where the two wave systems cancelled one another depended on the distance from the two openings and on the wavelength of the waves, Young was able to determine the wavelength of light from his experiment. He was able to show that for red light this was slightly less than one thousandth of a millimetre, and that the wavelength of violet light is only about half as great. It is therefore the wavelength of light that we experience as colour.

With this discovery, people understood how a prism affected light: it arranged the light of different colours intermingled in white light according to wavelength. Infrared radiation, like that discovered by Herschel, is light just like any other, but its wavelength is simply longer than that of red light. Continuing beyond both ends of the visible spectrum, we come to significantly longer or shorter wavelengths. Radio waves lie farther into the long-wave region. Relatively short, but still long in comparison with light, are radar waves, whose wavelengths are a few decimetres or centimetres. To take an example, the German radio station Bayerischer Rundfunk transmits in the medium waveband at 375 m. Long-wave stations transmit at kilometre wavelengths. On the short-wave side, the wavelengths of X-rays are only one-thousandth of those of light, while the wavelengths of gamma-rays are about one-thousandth of those of X-rays. Because electrical and magnetic fields play a decisive part in all these types of radiation, they are all known as forms of *electromagnetic radiation*.

WHAT IS LIGHT?

We have learned a lot about light, but despite this, we still do not know what light really is. The problem lies in the fact that it does not correspond to anything that we encounter in our everyday experience. We have already seen this in the apparent discrepancy between the particle and wave models. For physicists this is no problem. They are able to describe light formally using mathematics, and thus explain all its properties, and even predict how it will behave in a given experiment. Although we want to avoid mathematics in this book, we can obtain some assistance from a relatively simple model. We have already made use of one when we compared light to waves on the surface of water. The water-wave model is too simple. Light is more complicated than the waves produced when a stone is dropped into water. This is because light is not a mechanical phenomenon, but one where electricity and magnetism play significant parts.

Let us consider an ordinary wave of light as it passes us. To make it simple, we will imagine that we are watching its passage in slow motion. Let us suppose that we have a magnetic needle, like that found in any compass, and also an electrically charged body, say a small metal sphere. To give it an electrical charge, we could, for example, take a comb that had been electrically charged by combing someone's hair. If we were to bring the comb close enough to the sphere, we would hear the faint crackle of sparks jumping to the sphere, which would become charged with negative electricity. To preserve the charge on the sphere, we could fasten this to a silk thread that we can hold in one hand. Silk is a good insulator.

In darkness, both the electrically charged sphere and the compass needle lie motionless in our hands. Then a light-wave reaches us. Let us assume that we are looking towards it as it approaches us. We feel a slight tug on the hand that is holding the negative electrical charge. Whereas the charge on the end of the thread was previously perfectly still, it now suddenly pulls upwards, and we feel a slight force pulling on our hand. At the same time the magnetic needle swings to the horizontal, with the north pole pointing towards the left. Now the crest of the wave passes us. In the following trough the charge pulls downwards and the needle swings over until the north pole points towards the right. These events repeat themselves over and over: charge upwards, north pole to the left; charge downwards and north pole to the right. In Figure 3.6, the directions in which the negative charge tries to move and that indicated by the magnetic needle are shown schematically by arrows.

This was just a thought experiment. Even if we were to ignore the fact that the sphere and needle would move only slightly, the oscillations would

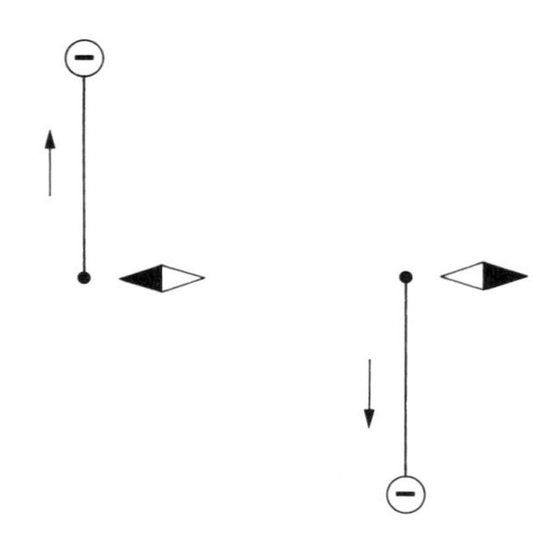

Figure 3.6. A thought experiment with an electrical charge and a magnetic needle. If a wave of light is regarded as coming towards the reader, i.e., at right-angles to the plane of the paper, at a specific moment the electrical charge would move upwards, and the north pole of the magnet would point towards the left (left-hand diagram). At the next instant, the charge would move downwards, and the needle would point towards the right

be so fast that it would be impossible for us to detect them: visible light causes electrons to move backwards and forwards 300 billion times every second. (Note that this is a European billion: one followed by 12 zeros or one American trillion!) In fact, light would not move a compass needle or start an electrically charged sphere swinging on the end of a thread. Light's electrical and magnetic fields are so weak that only the lightest electrically charged particles (electrons), react to it. Light falling on the surface of an object effects (at the most) just the electrons that are found in the body's atoms or molecules or, in the case of metals, that freely move around between the atoms. The light that we see actually causes chemical reactions in the rods and cones, the two types of cell in our retina, which in turn produce signals in the nerves leading to our brain.

From this experiment we can see that directional information is also implicit in light, and that this is apart from the direction in which it is propagating. We said that the charge moved upwards and downwards, and that the north pole pointed left or right respectively. There is also light where the charge moves towards the top right and bottom left, and the north pole points towards the top left or the bottom right. Depending on the incident light, the charge may move at any direction at right-angles to the direction of propagation. The direction in which the charge moves is known as the direction of polarisation. Our eyes are unable to detect the difference between light with different polarisations, but the polarisation of light is very important for many applications, including solar physics.

We asked what light really is. Light consists of an electrical field and a magnetic field. Both fields travel through space at the velocity of light, which means that they cover 300 000 km every second. The strengths of both fields vary like a wave, with crests and troughs. If such a wave of light encounters an electrical charge and a magnetic needle, both the latter will be set into oscillation at the same rate as the fluctuations in the fields that go fleeting past.

THE REVIVAL OF THE PARTICLE MODEL

We have now become accustomed to the wave model. It has explained why, if light is added to light, darkness may result, and also enabled us to measure the wavelengths involved—even though they may be extremely small. We now know that colour is nothing but wavelength.

Is Newton's particle model therefore dead? No, modern physics has revived it, recognising that light, and electromagnetic radiation in general, may occur in the form of tiny fragments, which are known as *quanta* (or *photons*).

The quantum model solves the old argument about the nature of light, and whether it consists of particles or waves. Light actually consists of innumerable individual flashes, or of waves emitted for a very short duration. These quanta are similar to Newton's 'corpuscles'. Each one, however, has a specific wavelength, which corresponds to the colour of the individual particles. These quanta travel through space in a straight line. When they encounter an object, they 'bump' into it, just as an individual seed would give a minuscule push when it hit anything. Light is able to move objects that it encounters. This so-called *light pressure* is extremely weak, at least for the strengths of light that we normally encounter.

But a quantum of light is also similar to a wave. If it is forced to pass through tiny openings, as in Young's experiment (Figures 3.3 and 3.5) it is found to have wave-like properties. It exhibits interference phenomena, similar to those shown by waves on the surface of water, and the wavelengths of the quanta may be measured.

Quanta are, so to speak, atoms of light. We are able to detect the quantum structure of light only in very difficult experiments. Normally it appears to us as a continuous stream. If we stand in the sunlight, we are not, however, bathed in a uniform stream of light. Instead, ten thousand billion quanta of sunlight fall on every square centimetre of our skin every second. That's ten thousand million million per second!

LIGHT FROM LUMINOUS BODIES

In a hot candle flame, tiny glowing particles of soot emit light, and the glowing element in a lamp bulb is a thin hot wire. On the Sun and the surfaces of stars hot gases are radiating the light. There is a simple law that applies here, and which is very familiar to us in everyday life. We know that a piece of iron that is just beginning to glow—such as the hotplate on a cooker—appears red. As it becomes hotter, its colour shifts towards orange. We use the term 'white-hot' for something that is particularly strongly heated. This is because then blue light—in other words, of even shorter wavelength—is also emitted, and the mixture of colours appears white to our eyes. The colour of a luminous body gives us a clue to its temperature. In general, another rule also applies: the higher the temperature of a luminous body, the more energy it emits. Every square centimetre of the surface of a hot star radiates more energy at every wavelength than the same area of the surface of a cool star. At the same time, because the temperature is higher, the maximum radiation shifts towards shorter wavelengths. It is for this reason that cool stars appear red, and hotter ones seem white or blue.

We are therefore able to deduce directly something about the temperatures of stars' luminous surfaces from the light that they emit. We know, for example, that the layers that emit the light that reaches us from the Sun have a temperature of 5500°C.

If we were able to look deep into the interior of the Sun, we would see material at a temperature of millions of degrees. Although the radiation from material at such temperatures would appear white to our eyes, in fact most of its energy lies in the X-ray region, which is invisible to us. There are, in fact, stars whose surface material is at a temperature of millions of degrees. Such stars may often be detected with telescopes that are specially designed to capture X-rays.

THE SOLAR SPECTRUM

The spectrum tells us far more, however, about the star that is our Sun. The deeper, and thus hotter, layers radiate over a wide range of wavelengths, and a somewhat cooler atmosphere lies above them.

Atoms of material have a property that has proved extremely useful to physicists: they absorb light at specific wavelengths. If, for example, we pass light though a layer of hydrogen gas and then examine the spectrum, we will notice that there is something wrong at a specific wavelength in the red region of the spectrum. There is a gaping hole in the spectrum of the light that has passed though the gaseous hydrogen: we see a dark 'line' (Figure 3.7). This is at a wavelength of 6.6 ten-thousandths of a millimetre (660 nanometres), and is known as the *hydrogen-alpha line*. Where is the light that is missing from this part of the solar spectrum? It is not lost. If we were to look from one side at the hydrogen gas that is absorbing light from the luminous source and see it against a dark background, we would find that it was glowing red. Seen from the side, the spectrum of the gas would show a bright line at just the

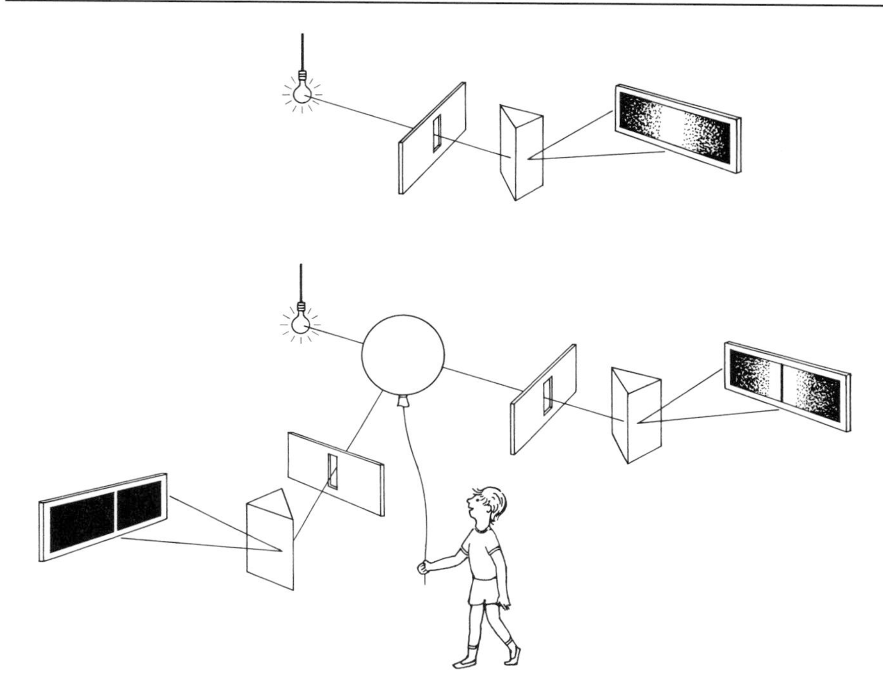

Figure 3.7. Top: light from a lamp bulb passes through a slit and then a prism producing a spectrum, with no visible spectral lines, and running from the violet (on the left) to the red region (on the right). This is known as a *continuous spectrum*. Bottom: if the light from the lamp passes through a gas—here shown as a transparent balloon filled with hydrogen—then the gas filters out light at certain specific wavelengths. A dark *absorption line* occurs in the continuous spectrum (right). The light that is absorbed is reradiated in other directions and produces a bright *emission line* in what is otherwise a dark spectrum (left)

wavelength at which the hydrogen had absorbed light. What has happened is that the hydrogen has absorbed light at one particular frequency from the radiation that would otherwise have travelled straight through the gas, and reradiated it in all directions.

Hydrogen does not just radiate and absorb red H-alpha light: there are an extremely large number of wavelengths at which it has similar properties. All atoms, not just hydrogen, have characteristic wavelengths at which they absorb and preferentially reradiate light .

Newton, in 1672, had spread out light according to its colour, and thus its wavelength. We know that the separation of the individual colours would have been better if he had used a slit, instead of a round hole in his shutters. Later, people discovered how to achieve even finer resolution of the colours by using ever finer slits. In 1802, the English doctor, technician and astronomer William Hyde Wollaston (1766–1828) noted that light was missing at certain points in the solar spectrum. The spectral band showed fine dark lines. Later these were called the *Fraunhofer lines*. They include the red hydrogen-alpha line. We already know how it occurs. Hydrogen, which is very abundant in the atmosphere of the Sun, removes light of precisely this wavelength from the radiation that is produced in much deeper layers of the Sun.

The son of a Bavarian glassmaker, Joseph Fraunhofer (1787–1826), who was one of the greatest optical workers of his time, discovered the dark lines independently of Wollaston. He systematically investigated the solar spectrum with his newly developed instruments, and designated the strongest lines with the letters A to I. These designations are still widely used. The hydrogen-alpha line was given the letter C. He recognised, however, that the solar spectrum contains a very large number of dark lines, and he counted a total of 574.

Since then we have discovered how to achieve a far greater resolution in dispersing sunlight according to its wavelength. Initially, all that was available was the *spectroscope*, which could be used solely to observe sunlight visually. With the introduction of photography, *spectrographs* (Figure 3.8) were developed, in which the spectrum could be recorded on a photographic plate.

Nowadays, the 26 000 lines known in the solar spectrum give us information about nearly all the chemical elements (Figure 3.9). Each type of atom removes its own characteristic series of lines from the spectrum. Because no two elements have the same lines, spectral lines are like fingerprints, from

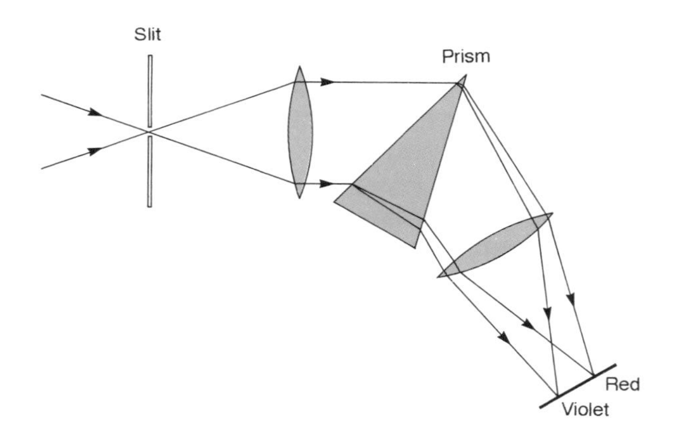

Figure 3.8. How a spectrograph operates. The light enters from the left through a small slit and is concentrated, as in Newton's experiment (shown in Figure 3.1), by a collecting lens and projected onto a prism, which disperses the rays of light into different directions according to their colour. A second lens collects the light dispersed by the prism and produces a band of colour, running from violet to red. If this is observed from the rear with an eyepiece, the instrument is a spectroscope, but if the spectrum is allowed to fall on a photographic plate, it becomes a spectrograph

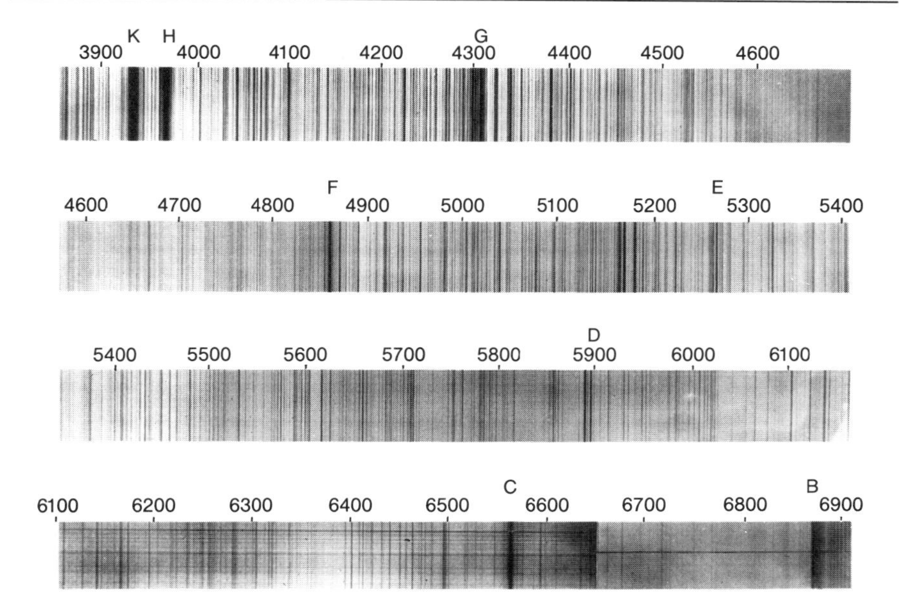

Figure 3.9. The spectrum of the disk of the Sun, divided into four sections, stretching from the violet (top left) to red (bottom right). The continuous spectrum is crossed by thousands of dark absorption lines. The figures are the wavelengths in ten-millionths of a millimetre. The letters are Fraunhofer's designation of spectral lines. The K line in the violet is a calcium line, which is discussed on p. 94. The C line in the red is the hydrogen-alpha line

which the individual types of atom may be deduced. This has given us the means of determining the chemical composition of the Sun.

But where is the light that has been removed from the solar spectrum at all these specific wavelengths? What happens in the atmosphere of the Sun is similar to the behaviour of the hydrogen gas that we held in the path of the light, as shown in Figure 3.7. When we look through the solar atmosphere at the Sun's disk, we see dark lines in the spectrum, just as we did when we were looking through the gas at a light-source. In explaining Figure 3.7, we mentioned that the light that was absorbed was reradiated in all directions, and that it is possible to see a bright line in the spectrum at the appropriate position when looking at the gas against a dark background. The same applies to the atmosphere of the Sun. If we observe the limb of the sun, particularly when all other light is blocked—as during a solar eclipse, when the Moon covers the rest of the Sun—we see a bright-line spectrum. Clouds of gas that lie above the surface of the Sun, known as *prominences*, which will frequently be mentioned later, show bright lines when they are observed at the solar limb against the dark sky background. When they appear in front of the solar disk, they are between us and the bright surface of the Sun, and their spectrum then shows dark lines.

The red hydrogen light plays a very minor part in the overall radiation emitted by the Sun. It is not noticeable to our eyes. We would need to have eyes that were blind to all other wavelengths and only sensitive to the red emitted by the hydrogen atoms. What would the solar disk look like to such eyes? Would it be evenly bright all over, or would it show different structures, which would be hidden to normal eyes? The answers to these questions came with the development of modern spectroscopic equipment.

4

TELESCOPES, SPECTRA, AND ECLIPSES

> . . . the Moon was directly in front of the Sun . . . what was visible around it was not simply the edge of the Sun, but an incredibly beautiful luminous ring, part bluish and part reddish in colour, split up into bright rays. It was just as if the Sun was pouring its stream of light down onto that from the Moon, and they were splashing out on top of one another.
>
> Adalbert Stifter, writing about the total eclipse of the Sun on 8 July, 1842

Our knowledge of the Sun would be insignificant, if we had not discovered how to extract more information from its light than that provided by a simple telescope or glass prism. In the next two chapters I intend to describe some of the milestones in the history of the development of modern equipment for observing the Sun.

I shall not keep to chronological order, but begin with the problem that still faces solar observers when they want to recognise individual features on the surface of the Sun. To do this they generally make use of the range of light that is detectable by the eye and ordinary photographic emulsions. Unlike research in just a narrow region of the spectrum, all the light, from red to violet, is used. This is known as 'white light'.

All astronomy began with white light, including study of the Sun. When Fabricius (both father and son), Scheiner, and Galileo examined the Sun with their telescopes, they were looking at sunspots in white light. The solar spectrum was still unknown. In 1859, Carrington—whom we met in Chapter 2—was observing the Sun in white light when he witnessed an explosion on the surface, which we shall discuss in Chapter 6. Although present-day solar researchers are able to extract a mass of information from the solar spectrum, they cannot forgo observations in white light. The 11-year increase and decrease in the numbers of sunspots, the multiplicity of shapes of spots and spot groups, and even the structure of individual spots themselves are all studied in white light.

Light spread out into a spectrum has told us which elements on the Sun are emitting light. Spectroscopic equipment was initially extremely primitive. Nowadays, giant evacuated tubes, situated at the bottom of solar towers, may stretch as much as 10 metres below the surface of the ground or into the face of a cliff. They contain instruments that are able to produce a spectrum, where the distance between the red and violet images is several tens of metres.

Any observer who uses just a narrow portion of the overall range of sunlight—say close to the red hydrogen line—sees a completely different (and unexpected) Sun from the one visible in white light. The stimulus that prompted people to develop this modern technique was actually provided by the Sun itself, and in fact arose in the few minutes when the Sun is invisible, hidden behind the Moon, during a *total solar eclipse*. I want to mention five eclipses that occurred in the last century, which not only produced new discoveries about the surface of the Sun, but which also ushered in a new epoch in solar physics. We will discuss the latter in full in the next chapter. First, however, let us take a look at the surface of the Sun in white light as seen by a modern solar observatory.

Even the best telescope will be of no use, however, if the sunlight reaching it has already been degraded as it has passed through the Earth's atmosphere. Even on a clear, starlit night, the light that reaches us from space is no longer in its original state.

THE STRUGGLE AGAINST TURBULENCE

The layers of the Earth's atmosphere do not just block out certain types of radiation, such as X-rays, but they also alter (in an uncontrollable fashion) the light that passes through them on its way to us. This is nothing to do with clouds, which may move across the Sun and the stars. Even when there is not a single cloud visible in the whole sky, the Earth's atmosphere may ruin the seeing. If astronomers did not have to breathe, they would gladly dispense with the Earth's shell of air.

When the layer of clouds clears as a depression passes, the stars and the Milky Way appear brilliant, and give the impression that they are nearer than they were on previous nights. Even so, the night may well be useless for astronomical measurements. Frequently when you look at a star through a telescope, all you see, instead of a small, brilliant, stationary point, is a patch of light, dancing around irregularly here, there and everywhere. The patch of light itself appears to be boiling and seething. The cause of this lies in irregular motions of the Earth's atmosphere. Rising and falling bubbles of air have various temperatures, and different densities. They therefore act like moving lenses, which continuously deflect light from a straight-line path. If we try to capture the boiling image of a star on a photographic plate, all we will obtain will be a fuzzy patch of light. The same problem makes it difficult to photograph features on the surface of the Sun. Unfortunately the disturbances in the Earth's atmosphere are never completely at rest. The possibilities for observing with ground-based instruments are correspondingly limited.

The irregular, unpredictable motion of a liquid or a gas is known as *turbulence*. Atmospheric turbulence is the bane of astronomers and the deadly enemy of any solar physicists who want to observe or photograph the disk of the Sun through a telescope. They see a wavering, unsharp image caused by the otherwise invisible inhomogeneities within the turbulent air. This is one of the reasons why solar observers send telescopes up to the top of the atmosphere with balloons, or launch them into orbit with rockets. Again, we shall be talking about this later, in Chapter 12.

Many of the findings that we have yet to talk about in this book have, however, been gained from the ground, and thus despite a constant fight with turbulence. Although we can never completely defeat our opponent, a suitable strategy will at least keep it within bounds.

Atmospheric turbulence is not equally strong everywhere. The image of the trees in a wood may, for example, wave about when seen through the layer of air above an asphalt road, but be perfectly still and sharp when seen across a lake, even in the midday heat. There is no turbulence above the lake, which is no hotter than the overlying air.

Anyone planning a solar observatory must therefore choose the site with care, which cannot be achieved without costly testing. This was why, in 1973, the German solar researcher Karl Otto Kiepenheuer (1910–1975), using an aircraft, measured the air temperature above the Roque de los Muchachos peak on La Palma in the Canary Islands to one hundredth of a degree. His instruments could even have detected temperature differences over distances of some 10 centimetres. Any turbulence would have been betrayed by small temperature fluctuations. The result was positive, with the atmosphere proving to be very quiet there. The effort expended in the careful investigation paid off. Today, a very successful Swedish solar observatory is situated on the mountain.

If we want to ensure that we are troubled by the least possible amount of turbulence, we have to locate observatories on high mountains, where sunlight passes through layers of air lying above a large expanse of water before reaching the telescopes. This is why there are now several observatories on the Canary Islands. As well as the Swedish telescope on La Palma, there are three German solar instruments on Tenerife, including the 38-metre high solar tower telescope. Sufficiently stable air is also found in Hawaii, where a solar observatory has been erected on Mauna Loa. The Sun is also successfully observed from the island of Maui, which is another of the Hawaiian Islands.

Even someone who is able to construct a solar observatory at the top of a high mountain on a small island in the middle of the ocean cannot be sure of avoiding turbulence. A turbulent layer of air forms above the ground when it is heated by the Sun. Anyone trying to avoid that needs to site their telescope at the top of a tower that rises above the agitated layers of air. The father of American solar physics, George Ellery Hale (1868–1938) built two solar towers, 20 and 50 metres high at Mount Wilson, north of Los Angeles, before the First World War. In the 1920s the Einstein Tower at the Astrophysical Observatory at Potsdam was constructed. This is renowned not just for the solar telescope that it contains, but also because it was designed by the famous architect Erich Mendelsohn (1887–1953).

In tower telescopes, the tower itself is generally used as the telescope tube. The telescope is therefore always 'looking' at the zenith. A *coelostat*, a large mirror at the top of the tower, initially captures the light and passes it, either directly, or via a second mirror, vertically down into the tower. The mirror (or mirrors) move so that the telescope continues to receive the light from the Sun as it crosses the sky during the course of a day.

Yet even the trick of using a tower is not sufficient to defeat turbulence. The building itself must be suitably designed. It is not surprising that astronomers are unable to see anything if they are forced to observe the Sun through the hot air rising from the flue of the administrative building's central-heating boiler. It is also obvious that the line of sight should not pass above a hot corrugated-iron roof. But the telescope tower itself should not heat up in sunlight, thus causing warm air to rise up the outside of the tower's wall, and ruin the seeing at the top. This is why all solar towers are now covered with a special, blindingly white, paint. The tower of the German Vacuum Tower Telescope on Teneriffe is surrounded by a layer of glass-fibre insulation, 10 centimetres thick, with a titanium-oxide paint that reflects all infrared radiation. Frequently the instruments are not surrounded by a wall at all and are supported by a skeleton framework, so the cool air has completely free passage from one side to another.

THE VACUUM TELESCOPE

When sunlight is collected by large mirrors and passed into the telescope tube for further investigation, the white light is, however, accompanied by infrared radiation, which heats the instrumentation. Warm parts of the equipment cause motion of the air, even if the currents are not as strong as those above a heated asphalt road. Turbulence dogs solar researchers even within the instruments themselves. To counter this problem, yet another trick has been adopted in recent decades. The air is removed from the telescope: without air there is no turbulence.

The Vacuum Tower Telescope at Kitt Peak Observatory, about 90 km from Tucson in Arizona, is an example of this. The tower is situated at the southwestern edge of the summit plateau. At the top of the tower, two mirrors reflect light from the Sun through a quartz window, 10 cm thick, and into the telescope tube. At the bottom the light falls on a concave mirror, 60 cm in

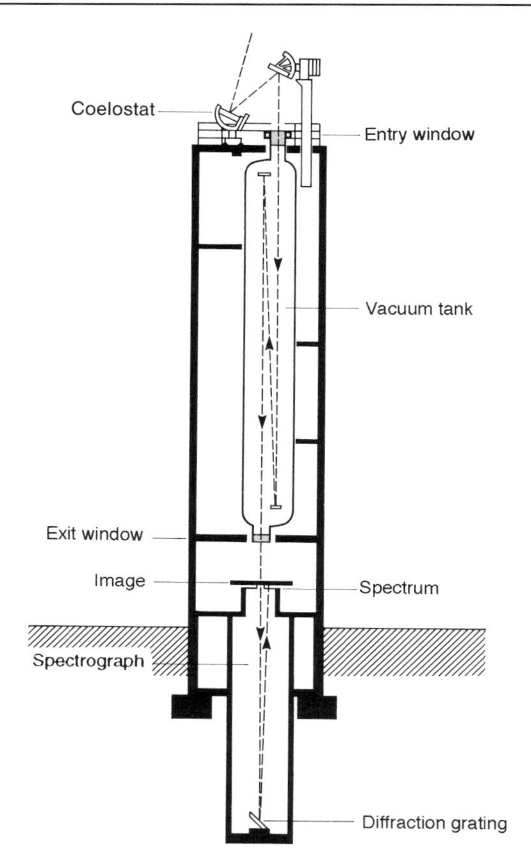

Figure 4.1. Diagram showing a vacuum telescope. A system of mirrors reflects the light down through the entry window into the vacuum tank. At the bottom of this evacuated tube it is reflected by a concave mirror back to a flat mirror at the top of the tube. From there, it returns to the bottom, where it leaves through the exit window. It then forms an image of the Sun, which may be studied by various methods, including, as shown here, by a spectrograph, which is buried in the ground. This spectrograph differs from the one shown schematically in Figure 3.8, in that it uses a diffraction grating rather than a glass prism

diameter, which reflects it back up to a flat mirror, from which it returns downwards, leaving the evacuated tube through a second quartz window. If a screen is placed in the light-path, an image of the Sun, 33 cm across, is produced (Figure 4.1). It is possible to see how the brightness of the Sun drops off towards the edge, and the umbra and penumbra of every spot may be distinguished. Additional investigations may be undertaken, by using a spectrograph, for example.

Modern telescopes have enabled us to see considerable detail on the surface of the Sun. The illustrations in Figures 2.5 and 4.2 were obtained with the Newton Vacuum Tower Telescope, operated by the Kiepenheuer Institute in Freiburg, Germany.

THE GRANULATED SURFACE OF THE SUN

Scheiner himself wrote that the surface of the Sun appeared to be rippled. In fact, small bright patches appear and disappear inside fainter surrounding material. The observer, who already has to contend with motion of the solar image caused by turbulence, is presented with a constantly changing picture.

In 1887, the French solar researcher Jules Janssen (1824–1907) began photographic monitoring of the solar surface. The quality of his pictures of the *granulation*, as the constantly changing, granular structure of the solar disk is called, was surpassed only in the 1960s.

Nowadays, using various means, it is possible to overcome the continual presence of turbulence and to capture clear pictures of the granulation (Figure 4.2). So we now know more about its nature. It consists of bubbles of gas that generally rise and sink with velocities of somewhat less than one kilometre per second. The diameter of these cells is about 1500 km—roughly the same as the distance between Munich and Madrid, or between Washington and Miami. If individual pictures are assembled to give a cine film, the boiling motion may be seen in slow-motion.

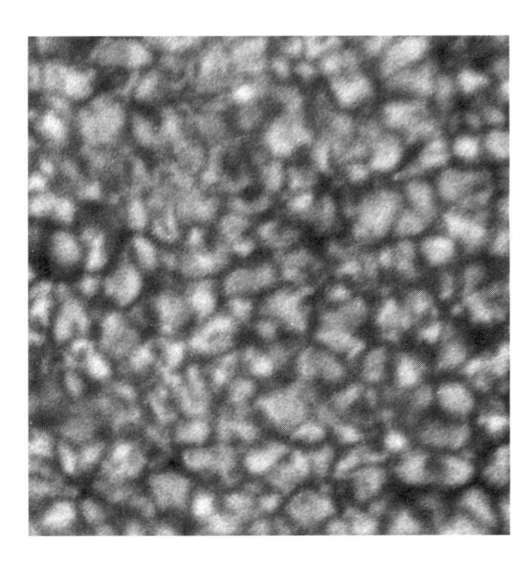

Figure 4.2. The solar granulation. Note that the bright granules are by no means circular, but generally angular in shape (photo: H. Wöhl, Kiepenheuer-Institut für Sonnenphysik, Freiburg)

The granulation is caused by the same effect that gives rise to the motion of air above an asphalt road: within the outer layers of the hot, gaseous Sun, heated bubbles of gas rise to the surface while other, cooler ones sink back into the interior, where they are reheated.

SUNSPOTS IN CLOSE-UP

Just as we examine something with a microscope if we want to detect more details than can be shown with a magnifying glass, so have modern telescopes revealed more details of sunspots, including features that were unsuspected in the last century. Figure 4.3 shows an enlarged picture of a spot. The exposure time was chosen so that the umbra, the least luminous, inner portion of the spot, appears black. In reality, it emits a blindingly bright light, although only one fifth of the energy of a similarly sized area of the undisturbed surface of the Sun. This is not surprising, given that the temperature of the umbra is about 2000°C lower.

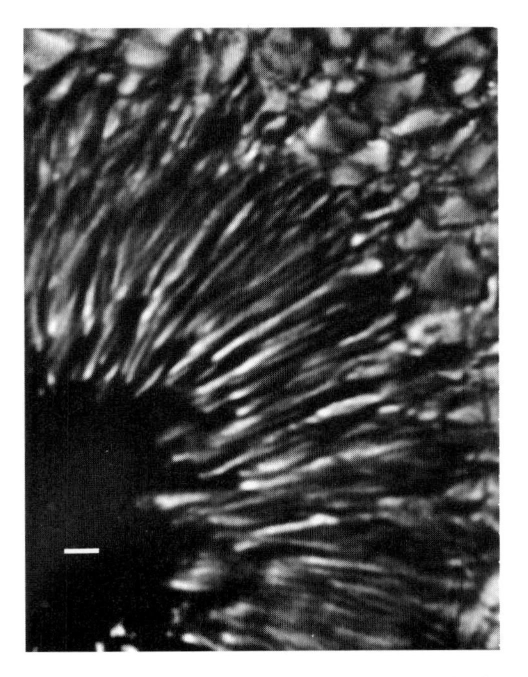

Figure 4.3. A detailed photograph of a sunspot, taken with the Swedish Vacuum Tower Telescope on La Palma. The umbra is bottom left, from which bright filaments run out across the penumbra. Outside the penumbra (top right) the luminous bubbles of gas that form the granulation are visible. The length of the white line shown for comparison purposes is 725 km, which is roughly the distance, as a crow flies, between Munich and Rome, or between Washington and Montreal (photo: G. Scharmer)

The penumbra exhibits fine structure, that was recognised in the last century. Long streaks run from the umbra right across the penumbra as far as the border with the undisturbed solar surface. Recently I saw a film taken by my Swedish colleague Göran Scharmer, using the vacuum telescope on La Palma. In slow-motion I could see a spot, surrounded by the boiling granulation of the solar surface. I could clearly see the narrow filamentary streaks that extended from the dark umbra out to the granulation. There were tiny bright points of light that moved inwards along these streaks, while waves of fainter luminosity streamed outwards along the filaments. It was a very impressive sight. Moreover, the apparently black umbra is by no means dead. With appropriate exposure times, photographs show bright points that appear and disappear within the umbra.

Telescopes reveal certain areas to be radiating brilliantly and others that are less luminous, to such an extent that they appear black by comparison. What materials are involved? Do the umbra and penumbra of a sunspot consist of different gases, and is the latter different again from the undisturbed surface?

THE SOLAR MATERIAL

The 'white' light that our eyes receive from the Sun reveals a picture of the solar surface with its granulation and sunspots. More information is concealed, however, within the light from the various brighter and fainter masses of gas. Each individual colour carries specific information. When the colours are mixed together to give white light, much of the information is lost.* Anyone who wants to find out more about the Sun abandons white light and turns to a spectrum.

In Newton's experiment, sunlight passed through a slit in a shutter and fell onto a glass prism. In both the last century and more recently, instruments were developed that dispersed light by passing it through a fine slit—generally only a fraction of a millimetre in width—and then through a series of prisms, mounted one behind another. Each individual prism amplified the effect of the previous one, until the light that had passed through the slit is spread out into a long band, with violet at one end and red at the other.

Prisms are rarely used nowadays to disperse light as a function of its wavelength. Reflecting surfaces covered in fine ruled lines will reflect light of various wavelengths in different directions. The colours that you can see on a CD disk are produced in this manner. Such mirrors, which are covered in extremely fine lines and disperse light like a prism, are known as *diffraction gratings*.

In the solar tower shown in Figure 4.1, the image of the Sun is brought to a

* It is easy to appreciate that mixing causes information to be lost if you imagine putting a newspaper through a shredder and then mixing up all the strips.

focus on the slit of a spectrograph. In the German Vacuum Tower Telescope on Tenerife, the spectrograph and its cylindrical, evacuated tube reach a depth of 16 m beneath the surface of the ground. There the light is dispersed by a diffraction grating into its various wavelengths. The ruled lines on a grating are extremely close together, with sometimes as many as 600 fine, parallel lines per millimetre.

Nowadays thousands of dark absorption lines are known in the solar spectrum. Every type of atom has its own, characteristic system of lines. They act like a set of fingerprints in the spectrum, and enable us (on Earth) to identify the atoms in the Sun. It is not just possible to determine what material exists in the Sun. The strength of the lines also enables us to tell the abundance of any particular element. For more than 100 years, it has been possible to carry out a chemical analysis of the material in the Sun, despite its distance of 150 million kilometres.

As early as 1862, lines of hydrogen were identified in the solar spectrum. Fraunhofer's C and F lines, and many others, are caused by that particular element. A decade later, 14 chemical elements had been found in the Sun that were also known on Earth. In 1868, a line in the yellow region was noted that could not be linked to any known terrestrial element. Because the element appeared to exist only on the Sun, it was called *helium*, the solar element (after the Ancient Greek word for the Sun). In 1895, the British chemist, Sir William Ramsay (1852–1916) discovered a gas in terrestrial minerals that produced the same spectral line. The solar element proved to exist on Earth as well. By the turn of the century, spectra had been obtained that showed as many as 200 lines of carbon alone. By then, practically all the known chemical elements had been detected in the Sun: only gold, mercury, bismuth, antimony and arsenic could not be found.

Modern chemical analyses of the Sun provide us with even greater detail. In a kilogramme of solar material we find about 700 grammes of hydrogen. Helium comes next, with about 280 grammes. The remaining 20 grammes include all the remaining heavier atoms and, in particular, carbon and oxygen. Of the elements known on Earth, only five are missing from the Sun. That does not mean that they are actually lacking. Some are radioactive and therefore mostly decay, and others have no lines in the regions of the spectrum that are available for analysis. A few are probably so rare that essentially no traces of them are visible in the solar spectrum.

Even gold has since been found. Its presence is betrayed by a line in the blue region of the spectrum. At first sight the amount of bullion that the Sun has managed to hoard appears extremely small: for every thousand million hydrogen atoms there are just nine of gold. Even such a vanishingly small proportion should not be underestimated, however. Because of the vast amount of material in the Sun, that still makes one six-hundred-thousandth of the mass of the Earth. If it were possible to transport even a tiny fraction of the Sun's gold to Earth, the bottom would fall out of the gold market. Bankers and owners of gold mines need not worry, however. The Sun's hoard is far better protected than in any earthly vault.

There is no difference in the chemical composition of sunspots and the quiescent solar surface surrounding them. The different appearance is caused simply by the difference in temperature.

SCATTERED SUNLIGHT

Turbulence is not the only effect of the Earth's atmosphere that bothers astronomers. When the Sun rises above the eastern horizon every morning, the whole sky brightens. Even in the west, opposite to the Sun, it takes on its customary blue hue. This is because the air does not allow all rays of light to pass through it in a straight line: some are deviated. This is why, during the day, light reaches us from every part of the sky, not just from where the Sun happens to be. The Earth's atmosphere is said to 'scatter' sunlight. It primarily affects short-wavelength, blue light: red light is far less affected. The rising or setting Sun appears red. In fact, some of the blue fraction of its white light is scattered aside when the radiation's path through the atmosphere is very long. White light appears red when some of its blue component is removed.

Without scattering by the air, the sky would appear as dark during the day as it does at night. The disk of the Sun would be blindingly bright in the black sky. This is how it appears to astronauts who look at it from outside the Earth's atmosphere.

The bright daytime sky not only swamps light from the stars, but also hides the relatively faint, luminous region surrounding the Sun. If we want to detect what is happening in the Sun's immediate vicinity, then two conditions need to be met. At first sight these appear to be contradictory, because first the Sun must be above the horizon (i.e., it must be daylight), and second, its light should not illuminate the Earth's atmosphere. This is only possible when the Moon hides the Sun's disk during the day, that is, during a total solar eclipse. Solar researchers have, in fact, learned a lot about the Sun in the few moments during which it is invisible.

WHEN THE SUN DISAPPEARS IN BROAD DAYLIGHT

The Earth orbits the Sun once a year, and during that time it is itself circled by the Moon about 12 times. To us, New Moon appears in approximately the same direction as the Sun. It is then invisible, moving in the daylight sky, generally passing above or below the Sun. Only occasionally does it hide part of the Sun's disk, which is when we observe a partial solar eclipse. Even more rarely, it passes directly in front of the Sun, and its central shadow cone falls on the surface of the Earth. At those places crossed by the Moon's shadow, night falls for a few minutes. The stars come out, flowers close their blooms, and astronomers open the shutters of their equipment.

Every 1000 years there are, on average, 659 total solar eclipses. The narrow tracks that are traversed by the shadow of the Moon are distributed over the whole surface of the Earth. The next solar eclipse visible from Germany will be on 11 August 1999. If it is raining that day, we will have to wait until 7 October 2135. Even with a much larger country such as the United States, there is a long interval between total eclipses. The last that crossed the USA, for example, was on 26 November 1979, and the next will be on 21 August 2017.

It is an eerie feeling when the Sun suddenly fades in the daytime sky and, seconds later, night falls. The event has always impressed spectators.

In the Old Testament book of Amos we find: 'On that day, says the Lord God, I will make the sun go down at noon and darken the earth in broad daylight.' Total eclipses were by no means unknown to the Hebrews. One could have been observed in August 831 BC over the southern edge of Palestine; there were others near Palestine in 824 and 763 BC.

Old accounts frequently describe total solar eclipses. When I first entered the monastic church at Weltenburg on the Danube in Bavaria, I was immediately struck by the altarpiece that hangs on the left-hand wall, near the entrance.

The St-Benedict altar was created by the two Asam brothers between 1734 and 1736. The altarpiece shows St Benedict, who is staring in amazement at the sky (Figure 4.4), where there is a dark disk, surrounded by a luminous halo. At bottom left, a ray of light streams out and falls on the eyes of the saint. Several experts whom I asked explained that while the rest of the monastery was asleep, St Benedict had climbed to the top of a tower, where, in a state of profound religious ecstasy, he had seen a vision. But this, they said, had nothing to do with the Sun. Despite this explanation, I was unable to get the idea out of my head that had come to me when I first set eyes on the painting: whatever its significance, the painter must, at some time in his life, have witnessed a total solar eclipse. He must have seen how the dark disk of the Moon appeared in front of the Sun, surrounded by the corona. He must also have seen how, at the end of the spectacle, the first ray of light broke through at the edge of the Moon, just as I had experienced in 1961.

If either of the two brothers Asam had seen a total solar eclipse, it could only have been the one of 12 May 1706. One of them was 20 years old, and the other 14. However, the eclipse was not visible as total in Bavaria. Anyone wanting to witness it would have had to be in northern Germany.

During the short time of a total solar eclipse, the Moon covers the Sun's disk, and it is possible to see the region immediately surrounding the Sun, which is normally hidden by the Sun's glare (Figure 4.5). From various points around the black disk of the Moon, there are red, unmoving tongues of flame that project into the dark sky, and which are known as *prominences*. The Sun

Figure 4.4. The St-Benedict altarpiece in the monastic church at Weltenburg. Was the painter inspired by a total solar eclipse?

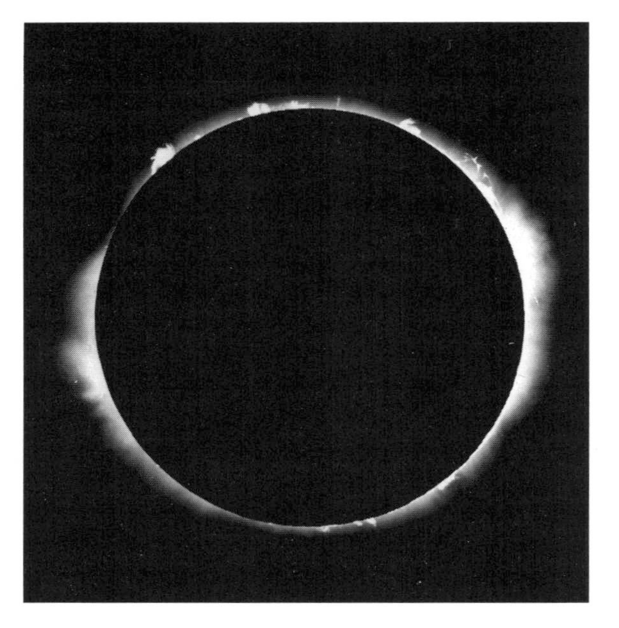

Figure 4.5. A total solar eclipse. The Moon's disk has intercepted the light from the Sun, and around its edge the sharply defined luminous 'tongues' of prominences may be seen. Also visible is some of the diffuse light from the corona, which does not show well with the short exposure time used here. A photograph of an eclipse with an exposure chosen to show the corona is given in Figure 6.1

is also surrounded by a halo of softly glowing white light, which extends far out into the dark sky. This is the *corona*. The event, which, for a short time, brings the whole of nature to a halt, lasts but a few minutes: never longer than eight minutes, and generally much less.

Total solar eclipses are therefore highly desirable events that solar researchers do not want to miss. At this very moment, preparations are under way for expeditions to ensure that instruments will be ready, at the right time, on the small strip of the Earth's surface that will be crossed by the path of the next eclipse.

FIVE DECISIVE ECLIPSES

In the last century, five total solar eclipses occurred within a period of 11 years that made a decisive contribution to our understanding of the nature of the various phenomena that were visible. One question that had remained unanswered until then was whether the prominences were a real phenomenon, or were simply an optical illusion. If they were real, then it was a question of whether they were part of the Moon or of the Sun.

The answer was given by the eclipse of 18 July 1860, over Spain. Two workers transported their instruments to the eclipse track, one to the Ebro, and the other to the Spanish Mediterranean coast. For the first time photography—then still known as Daguerreotyping—was employed. The Englishman, Warren de la Rue (1818–1887), a wealthy paper manufacturer, was a pioneer in the photography of astronomical objects. The Jesuit, Angelo Secchi (1818–1887), Director of the Collegio Romano in Rome, devoted the second half of his life to the study of the Sun. Both men photographed the prominences from points in Spain that were several hundred kilometres apart. The pictures agreed with one another like two peas in a pod. The red tongues of flame were therefore certainly real. Even more, plates taken rapidly one after the other showed that the Moon moved *in front of* the prominences, so the latter were part of the Sun.

Eight years later, on 18 August 1868, the Moon's shadow crossed the Indian and Malaysian peninsulas. This event ushered in the greatest discovery of modern solar physics, to which we shall return in the next chapter.

The eclipse of 7 August 1869 could be observed along a narrow track that stretched from the Behring Strait right across North America to North Carolina. This time the corona was tackled. It was not easy to obtain the spectrum of this faintly luminous veil of light. At first there seemed to be nothing startling about it. The light appeared to be spread evenly across the whole spectrum. The dark Fraunhofer lines in the solar spectrum were missing. There was, however, a bright line in the green. What atom was responsible? Precise measurements showed that no known atom emitted light at that wavelength. Was this evidence for the existence of an element that was unknown on Earth? Towards the end of the last century, the name 'coronium' was given to this unknown material. This puzzling element never appeared to be present in the Sun itself, because there was no Fraunhofer line to betray its presence. It was not until the middle of this century that the secret of the green coronal line was revealed. It was produced by atoms of iron. In the corona, the gaseous iron is so hot that the atomic nuclei, which are normally surrounded by a cloud of 26 electrons, have lost 13 of them, leaving just 13 to orbit the nucleus. This can occur only at extremely high temperatures. We now know that the solar corona has temperatures of millions of degrees.

The fourth of the eclipses was short. On 22 December 1870, the Sun was hidden behind the Moon's disk for just two minutes and ten seconds. Anyone who wanted to observe this event had to travel to the Mediterranean, which was not always easy. Janssen, in Paris, had built a reflector specially designed for the light of the corona. But Paris was under siege by the Prussian army. So he arranged for himself and his equipment to be flown over the Prussian lines in a balloon. Despite this, he had bad luck. Oran, where he intended to observe, was covered by a thick layer of cloud. Sir Norman Lockyer (1836–1920) fared hardly any better. He travelled to Sicily, but the *Psyche*, on which he had embarked, was shipwrecked. In the end, Lockyer was actually able to see the eclipsed Sun—for one and a half seconds!

Charles August Young (1834–1908), who had travelled from Princeton University in New Jersey, where he was a professor, had better luck. As the Moon covered more and more of the Sun, Young observed the spectrum of the Sun. He saw the Fraunhofer lines becoming fainter and fainter, as less light reached the spectroscope. Young himself wrote: '... the dark lines of the spectrum, and the spectrum itself gradually faded away: until all at once, as suddenly as a bursting rocket shoots out its stars, the whole field of view was filled with bright lines more numerous than one could count. The phenomenon was so sudden, so unexpected, and so wonderfully beautiful, as to force an involuntary exclamation.' The spectacle lasted just two seconds. Young was fairly confident that for that short space of time the solar spectrum was reversed. What had previously been bright, i.e., the continuous background, became dark. What had previously been dark, i.e., the Fraunhofer lines, suddenly became luminous, line for line.

Since then, the phenomenon has been observed at every total solar eclipse. Shortly before the Sun disappears behind the Moon, it appears as a narrow crescent that becomes smaller and smaller. At the very last moment, we perceive light from just the uppermost layers. Remarkable changes in the spectrum of the Sun then occur. No slit is required in spectrographs, because, shortly before total eclipse, the Moon leaves just a narrow, semicircular strip uncovered. In the spectrum the white light, which is evenly spread over all wavelengths, disappears. Instead, bright lines are present. The layer of the Sun from which this light originates is known as the *chromosphere*.

Readers will recall from our discussion of Figure 3.7, why the Fraunhofer lines shine in emission shortly before the end of totality. The atoms in the solar atmosphere, which filter out light at the wavelengths of the Fraunhofer lines, reradiate it in all directions. When we look at the edge of the Sun our line of sight grazes the limb and we see the solar atmosphere against the dark sky background, and therefore see only the light radiated by the atoms in a tangential direction, i.e., parallel to the solar surface. This light is at precisely those wavelengths at which the atoms have previously absorbed light, in other words, at the wavelengths of the Fraunhofer lines. Normally the light from the daytime sky overpowers the phenomenon, but it becomes apparent when the Moon almost completely covers the Sun, and thus reduces the brightness of the daytime sky.

The eclipse of 12 December 1871 could be observed from India and Australia. Janssen again joined in. He was successful in detecting, in the spectrum of the corona, where Young had found his bright green line, dark lines, just those lines that had been previously recognized in the spectrum of the solar disk. Janssen found, for example, Fraunhofer's D line in light from the corona. It seemed as if the corona was merely reflecting sunlight.

ECLIPSES WITHIN A TELESCOPE TUBE

The few minutes of a total solar eclipse are precious. Frequently expeditions that have been prepared over a period of years fail because of bad weather. For more than 100 years, therefore, it was astronomers' earnest wish to be able to observe, outside eclipses, the faintly luminous regions that exist above the surface of the Sun.

A method of achieving this for the corona was devised in the 1930s. We are forced to contend with light scattered in the Earth's atmosphere, but our view of the corona is also affected by light that enters the telescope and is scattered in all directions within it. Light reflected from the inner wall of the tube also affects observation. After 20 years of hard work, the brilliant French optician, Bernard Lyot (1897–1952) built a telescope in which the stray light was so greatly reduced that the corona became visible in daylight. Lyot noted that a large portion of the scattered light arose from the objective—the lens at the entrance to the telescope. Although the objectives of refractors normally consist of two individual lenses of different sorts of glass, cemented together, Lyot used a single lens, which produced far less scattered light. To do so he had to use glass that was free from bubbles and internal streaks (striae). The surfaces of the glass had to be extremely finely polished, so that there were no scratches to produce stray light inside the telescope, and the lens had to be kept free of dust particles. In Lyot's instrument, an objective lens produces an image of the solar disk on an inclined mirror, which directs the light away from the optical axis, where it may be dealt with harmlessly (Figure 4.6). The inclined mirror is of a size such that it intercepts just the image of the Sun, but

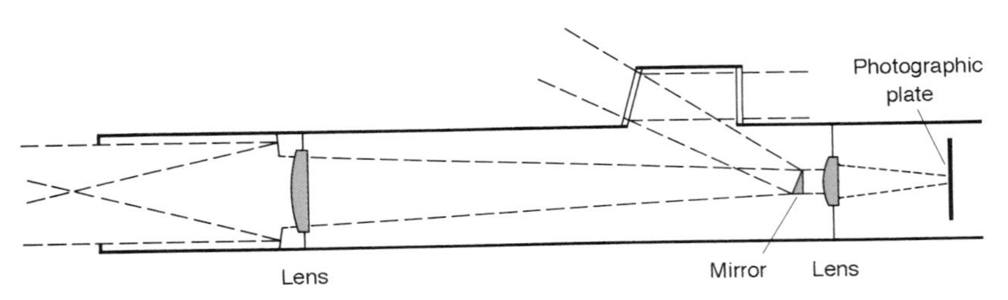

Figure 4.6. Schematic diagram of a coronagraph: sunlight enters from the left and falls onto a lens (the objective) producing an image of the Sun on the right. A round, inclined mirror diverts the glare from the Sun's disk into a side chamber, where it is harmlessly dispersed. Light from the corona surrounding the solar disk, bypasses the mirror, however, and falls on a second lens, which produces an image on a photographic plate (right). In this image, the blinding light from the Sun is hidden. The faint glow of light from the corona therefore becomes visible even against the bright daytime sky

allows light from the surrounding region to pass by, to fall on a second lens. Any dust in the air within the tube itself can create unwanted scattering. The inside of the tube is therefore greased so that any dust is captured as if by fly-paper. Diaphragms intercept any remaining scattered light that, despite all the precautions, might otherwise affect the sharp image of the corona. An image is produced behind the second lens. This shows the black rear side of the intercepting mirror, around which the regions immediately surrounding the Sun may be seen, just as they can around the Moon's disk at totality. The circular mirror acts like the Moon.

Using this device, known as a 'coronagraph', it has been possible to investigate the corona outside eclipses ever since the 1930s. Even more success has been had in observing prominences. All this began with the eclipse of 1868.

Among the five notable eclipses in the middle of the last century, we have only briefly mentioned the Indian eclipse of 1868. There is a lot to say about it, because it ushered in a new era in solar research. This required considerable time however. The man who made the greatest contribution was just seven weeks old at the time of the Indian eclipse.

5

THE MONOCHROMATIC SUN

> Behind an old trunk in the attic he discovered some lusters from a candelabrum. He washed them, split one of them . . . and mounted it on a tin stand. From a cardboard tube and a pair of old spectacle lenses he made a collimating tube and painted it black inside. Into this tube he fitted another tube "like a sword into its sheath"; into this second tube he fitted a slit made from a brass plate. . . . In that night he looked through his eyepiece at a candle flame . . . the continuous spectrum shone out, with the red, green, blue and violet . . .
>
> Helen Wright, writing about the thirteen-year-old George Ellery Hale in *Explorer of the Universe* (1966)

What happens inside a spectrograph? Newton himself needed a lens as well as a glass prism to help produce a spectrum. Like Newton's arrangement, modern spectrographs still have a narrow slit, through which the light enters. Following that, it passes through a lens and falls onto a prism. This was shown schematically in Figure 3.8: light falling on the slit from different directions is collected by the lens and directed onto the prism. There it is spread out according to its wavelength. If it were not for the fact that the prism disperses light of different colours in different directions, then the second lens would produce an image of the slit as a narrow white line. However, the prism disperses different wavelengths to different degrees. It therefore produces an image of the slit for every individual colour. A spectrum is therefore a series of adjacent images of one and the same slit, each of which is in a slightly different colour. Figure 3.8 shows how the image of the slit at red and violet wavelengths is formed at different points in the spectrum. As we have already seen, in modern large solar telescopes, prisms have been replaced by reflecting gratings: mirrors that reflect light in different directions according to wavelength.

If the simple spectroscope shown in Figure 3.8 is turned towards the Sun, it would disperse light from a wide region of the spectrum, in fact all the light that passed through the slit and fell on the first lens. Light from the whole solar disk would contribute to a single spectrum. Using such a method it would never be possible to discover how the light from the centre of a sunspot (say) differs from that of the quiescent surrounding solar surface. To do so, it is essential to allow just light from the centre of the spot, and no other, to enter the instrument. But if the spectroscope is combined with a telescope, that presents no problem.

We need to carry out a thought experiment, in which we repeat Newton's experiment with the prism and the slit in the shutters, but vary it slightly by using a telescope, with which we, like Scheiner and Hevelius, project an image of the Sun onto a screen. We have already discussed the projection method in Figure 2.2. It is the same as that described in Appendix A for the observation of the Sun with binoculars. To understand the principles behind the telescope/spectroscope combination, we can imagine that Newton and Hevelius combined their experiments.

ONCE UPON A TIME, NEWTON VISITED DANZIG

In fact the two never met. Let us, however, give our imaginations free rein, and say that in 1670, the 27-year-old Isaac Newton visited Danzig. By then he had already carried out his experiments with prisms. Let us assume that he had brought a prism in his luggage, and met Hevelius, who at 59, was the most famous person in Danzig, which at that time had about 60 000 inhabitants. Let us also imagine that Hevelius showed him the Sun, using the projection method.

When Newton saw the bright image of the Sun on the sheet of paper, he had an idea. He found his prism, and asked Hevelius to cut a narrow slit in the sheet of paper. The latter agreed, and the young Newton repeated his earlier experiment, which we have described on page 34. This time, however, instead of the slit being in the shutter, it was in the paper screen on which Hevelius was projecting the image of the Sun. The lens and prism produced a spectrum on a second screen (Figure 5.1). The light was no longer derived from the whole of the solar disk, but from just the narrow slit in the paper screen. Light from each point of the slit came from a specific point on the Sun. The broad strip of the spectrum consisted of innumerable narrow individual spectra. If Hevelius arranged his telescope so that the slit lay across a sunspot, there would be a somewhat fainter strip running along the length of the spectrum. This was the spectrum of the sunspot, bordered on both sides by the spectra of the normal, quiescent solar surface.

Let's extend our fantasy about these two scientists having met, and assume that both the telescope and spectroscope were good enough to meet modern demands. What would the two have seen?

Thousands of dark Fraunhofer lines were visible in the bright spectrum on Newton's screen. Both men were able to recognise the H-alpha line in the red, for instance, which appeared as a dark transverse line, where

Figure 5.1. The projection principle, as used by Hevelius, and shown in Figure 2.2 bottom, combined with Newton's experiment of Figure 3.1. The spectrum obtained is not that of the whole solar disk, but of the small portion of the image produced by the telescope that falls on the slit. Note that this principle is used in the tower telescope shown in Figure 4.1. The image produced by the telescope falls onto a screen, where a spectrograph slit allows the light from just a narrow strip of the image to pass

atoms of hydrogen absorbed light and reradiated it in all directions. When the slit in Hevelius' slit crossed a point where a cloud of cooler hydrogen gas was suspended above the surface of the Sun, they saw that the dark line was stronger, because the gas absorbed even more light. At the limb, however, where normally one would be looking at the background sky, they saw something remarkable. Wherever a cloud hovered above the surface of the Sun, it appeared alongside the solar disk. The two men then saw the normally dark line appear bright in front of the dark sky background. They were seeing the light that the hydrogen had absorbed and was reradiating in all directions. The line that they saw was dark in front of the Sun, and bright in front of the dark background. The two men saw prominences without a solar eclipse.

Regrettably, Newton and Hevelius never carried out any experiments together. It was hundreds of years before it was possible to observe prominences daily. The principle described in our tale, however, is still employed today, even though the solar image is not projected in the same way as by Hevelius. The objective lens or the primary mirror in a telescope itself produces such an image of the Sun inside the telescope. We have seen that the solar image produced by the Vacuum Tower Telescope at Kitt Peak has a diameter of

330 mm. The image may be intercepted with a screen. If this screen has a slit, which is also the entrance slit to a modern spectrograph, then we have essentially the same situation as the thought experiment that we have just carried out in our alternative history. Nowadays, solar researchers are not only able to obtain images of the Sun in just the light of a single line, but are able to capture processes occurring on the Sun on film and at individual wavelengths. They can also draw up 'maps' of the velocities involved in eruptions, and even daily magnetic maps of the Sun. We shall discuss all of these later.

The total solar eclipse of 1868 played a key part in this. The decisive step, however, was made by the American, George Ellery Hale (1868–1938).

THE YOUNG MAN WHO BECAME A SOLAR RESEARCHER

After both of her first two children had died as infants, neither Mrs Hale, nor her husband William, expected their third child, a boy, to live long. Both he and his younger sister survived the usual childhood illnesses, however. The Hales were then living in the centre of Chicago, but roughly two years after George Ellery's birth, they moved into one of the suburbs. They thus escaped the flames of the Great Fire of Chicago of 1871, which totally destroyed the house in which George had been born.

About this time, Hale the elder founded a firm manufacturing elevators. With the ever-increasing height of buildings in American cities, the company had no lack of orders. It also did well in Europe. Later it even constructed the elevators for the Eiffel Tower in Paris. In 1886, Hale the elder made a business trip to Europe. In private, he hoped that his son, then 18, would follow him into the firm.

At first the boy had been interested in microscopy, then he built himself a chemical laboratory. Reading Jules Verne's novel *From the Earth to the Moon* impressed him so much that he called his laboratory an 'observatory' and became increasingly interested in astronomy. His hobby remained subordinate to the elevator business, and the young man was technically skilled and properly trained. His father was still able to hope that he would eventually follow his wishes.

At home, Hale had carried out a lot of experiments with the solar spectrum. At 17 he had measured the Fraunhofer lines with his own spectroscope. His interest in the solar spectrum had already brought him in contact with Samuel P. Langley (1834–1906), Director of the Allegheny Observatory in West Virginia, who was able to give him some useful tips and advice for his experiments. Hale had also read two books by the English solar physicist, Norman Lockyer, and he regretted that he had had no opportunity during his European trip to talk to Lockyer in person. His visit to Jules Janssen, the Director of the Meudon Observatory near Paris, however, was compensation. Independently of one another, Janssen and Lockyer had made a great discovery in 1868, the very year of Hale's birth.

THE INDIAN ECLIPSE

In 1868, Janssen travelled to Guntur in India for the total solar eclipse. Using his spectroscope during the short period of totality, he noted that the prominences radiated mainly at the wavelength of hydrogen light. When, after a very short time, the Sun reappeared from behind the Moon, overpowering the prominences, Janssen must have said to himself 'I want to see those lines outside eclipse.' The next day at sunset, he directed his telescope, which was fitted with a spectroscope, at the point on the solar limb where he had seen a particularly bright prominence. Janssen observed the Sun by a method similar to the one used by Hevelius and Newton in our fictional tale.

When Janssen looked at the hydrogen-alpha line through his spectroscope, he was essentially blind to all other light. With his narrow slit, he was looking at the Sun in light of just that wavelength. He was actually able to detect the spectral lines of the prominence even in the bright daylight sky. Through the slit of the spectroscope he could see, in the red light of the hydrogen line, a narrow strip of the luminous masses of gas that extended beyond the solar disk. If he moved the telescope slightly, then the slit moved across the image, and showed a neighbouring narrow strip of the prominence. Through the spectroscope, Janssen was thus able to examine one strip of the limb of the Sun after another, draw what he saw, and build up a picture of the whole prominence.

Work carried out during an eclipse is always hectic, because there are only a few minutes at the most, and there will probably not be another opportunity for years. Now, however, Janssen could draw prominences at leisure, using the spectroscope attached to his telescope. He could follow them, watching how, over a period of hours, they expanded and then sank back again, and how they dissipated or, because of the rotation of the Sun, disappeared over the limb after a few days. Janssen was so enthralled by this fascinating display that it was a month before he submitted a report about the discovery to the French Academy of Sciences. Five minutes before his letter reached the Academy, a paper by Lockyer was read, in which he described how he had succeeded in observing prominences outside an eclipse. Janssen had indeed been the first to observe prominences in daylight, but Lockyer published his findings five minutes earlier! The Academy's judgement was worthy of Solomon: they had a medal struck to commemorate the discovery: one side shows the portrait of Lockyer and the other that of Janssen.

I mention Janssen and Lockyer in particular, because Hale built on their work. Although both were able to see prominences when they extended beyond the solar limb and were thus projected against the dark sky, it was

Hale who succeeded in detecting prominences in front of the bright solar disk.

Apart from his meeting with Janssen, the most exciting event of his European trip for Hale occurred in London: in a scientific-instrument shop he was able to purchase a good-quality spectroscope.

HALE DEVELOPS THE SPECTROHELIOGRAPH

The Janssen–Lockyer method of observing prominences through the slit of a spectroscope attached to a telescope ushered in a new era in solar research. It was possible to observe how prominences appeared, how they frequently reached their maximum brightness within an hour, rose to heights of hundreds of thousands of kilometres, disintegrated into individual filaments and faded away. The solar researcher, Charles Augustus Young, whom the young Hale visited at Princeton, New Jersey—we have already met him in connection with the eclipse of 1870—succeeded in photographing a prominence through a slightly widened spectroscope slit. But it was not a good picture. One cannot widen the slit of a spectroscope with impunity. No details could be seen in the exposure. Hale, who was then just 21, brooded on this, and how it might be possible to obtain better photographs despite interference from the light of the Sun, and thus be able to record more on the plate.

Later Hale said that the idea came to him when he was sitting in a tram, looking at a picket fence and the garden that was visible beyond it through the gaps between the palings. The inspiration for the *spectroheliograph*, as Hale called the instrument, came into his head at that moment.

With hindsight, the idea appears an extremely simple concept. A telescope produces an image of a prominence in the same plane as the spectroscope slit; the slit selects a narrow strip of that image, and just the light of that small strip passes into the spectrograph. This then produces an image, on the photographic plate, of the slit in all colours. We can now place a second slit at the point on the spectrum where the hydrogen-alpha line occurs. This second slit passes just the light from the hydrogen line that originates in the section of the solar disk selected by the first slit. A photographic plate behind the second slit will therefore receive an image in which the first slit has selected a narrow section of the solar disk, and the second slit a narrow region of the spectrum (Figure 5.2). If the plate is developed, it shows an image of the prominence in light at the selected wavelength. However, the image will show just a narrow strip of the object, and, rather like looking at a picture by Rembrandt in an art gallery through the narrow crack of a door from a neighbouring room some distance away, hardly anything will be seen.

Hale overcame this with a very neat method. The telescope is moved slowly relative to the Sun, causing the image of the prominence to move gradually across the slit. To put it another way: the small strip to be photographed slowly

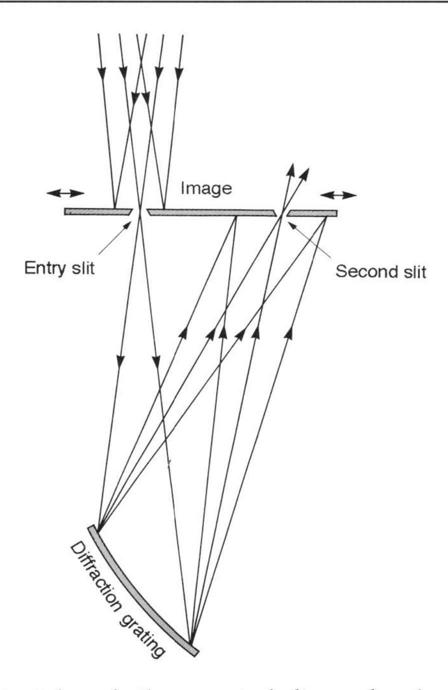

Figure 5.2. The principle of the spectroheliograph. As in our imaginary Newton/Hevelius experiment of Figure 5.1, a small section of the image at the focal plane of the instrument is selected with a narrow slit. The light is reflected upwards from a diffraction grating, producing a spectrum, again at the focal plane. Here, a second slit selects light of a specific wavelength from the spectrum, e.g., light corresponding to the red hydrogen line. If the plate containing the two slits is moved horizontally, then the entrance slit passes adjacent strips of the image. Simultaneously, the light leaving through the moving exit slit builds up an image on a photographic plate (not shown here), thus producing an image of the Sun at a specific wavelength

crosses the prominence. If the photographic plate behind the second slit is also moved at the correct speed, then the neighbouring strips are captured on the plate. An image of the whole prominence is thus produced on the emulsion. At the time, Hale did not suspect that other solar researchers had had the same idea previously. For example, Oswald Lohse (1845–1915) from the Astrophysics Observatory in Potsdam had already described the principle in an earlier paper.

Hale studied at the Massachusetts Institute of Technology. There, he succeeded in getting his supervisor to agree to his taking 'The Photography of Prominences' as the topic for his thesis. He thus managed to combine his studies at the Institute with the work that he carried out both in his private observatory at home and at Harvard Observatory.

For a long time he had no success. He was finally lucky on 14 April 1890.

'On April 14 a cool breeze was blowing, making the seeing fair in spite of a little whiteness in the sky', he wrote later about that day, 'A hasty examination of the limb discovered a prominence in a good position for the work, and a photograph was made through F, the slit being about 0.0005 inch wide. On developing the plate, the outlines of two prominences could be seen rising above the limb. As only one prominence had been noticed in observing the point in question, I returned to the telescope, and found that there were in fact two prominences in the exact position shown in the photograph.' Later, one of Hale's friends said that the discovery had undoubtedly added more to our understanding of the processes occurring in the sky than any other since Galileo first turned his telescope onto the heavens.

A year later, Hale was again in Europe. The 23-year-old was now well known in the field of solar research. He was invited to meetings and had the opportunity to meet and talk to notable astronomers and physicists. In London, he came across, by accident, a French scientific publication in which Henri Alexandre Deslandres (1853–1948) of the Observatory of Meudon near Paris had published the principle of the spectroheliograph. Deslandres—like Hale and Lohse—had independently thought of the same idea. However, Hale's contribution as the first to apply the principle in practice remained unchallenged.

Following his return to America, he succeeded in recording the whole of the solar limb and all its prominences on a single plate. The time when observers had to observe prominences laboriously through a narrow slit was, thanks to Hale's discovery, completely over.

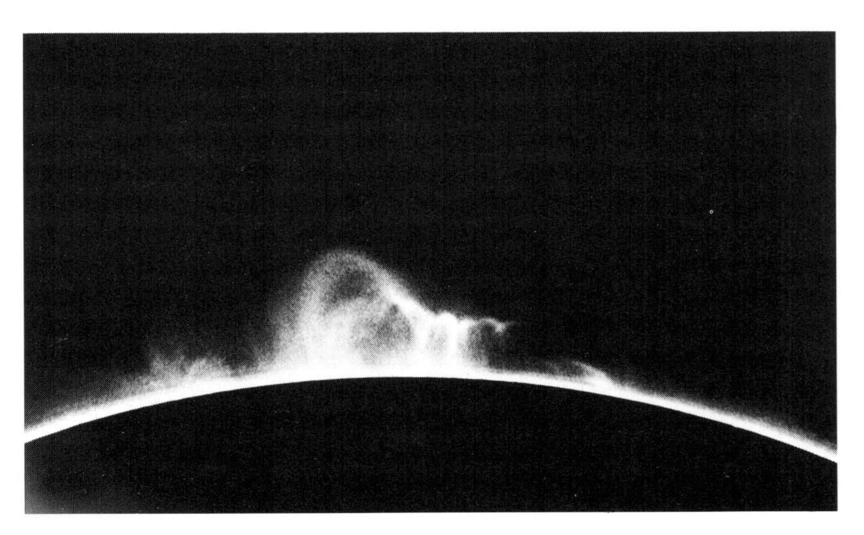

Figure 5.3. Solar prominences, photographed in the light of the red hydrogen line, soar above the edge of the solar disk, which has been artificially occulted (photo: High Altitude Observatory, Boulder, Colorado)

Hale's method of observing the Sun in a single colour, however, did not just enable one to see prominences at the solar limb, where they emitted light in numerous spectral lines and were visible against the bright sky caused by scattered sunlight. Prominences were also visible when, as seen from our viewpoint, they appeared against the solar disk.

Thanks to a discovery by the French astronomer, Bernard Lyot, a spectroheliograph is not required nowadays to observe prominences. It is possible to manufacture filters that pass light of a specific wavelength only. With a Lyot (interference) filter that passes light of just the hydrogen-alpha line, glowing prominences may be seen at the solar limb, because, as in a spectroheliograph, light at all other wavelengths is rejected. Using such an interference filter it is possible to photograph the Sun in light of that wavelength. We thus obtain a picture that shows us where hydrogen at the limb is radiating at that wavelength (Figure 5.5). Hydrogen clouds may be seen in front of the disk as well, when they appear dark in hydrogen-alpha light (Figure 5.4) and are known as *filaments*.

Figure 5.4. The Sun in hydrogen-alpha light. Near sunspots hydrogen is particularly bright. The dark objects on the solar disk are filaments, i.e., masses of gas suspended above the surface of the Sun. Their hydrogen absorbs light at the hydrogen-alpha line (which is why they appear dark) and reradiates it in all directions. When seen at the limb they therefore appear bright, as in Figure 5.3, and are then known as prominences (photo: Haleakala Observatory, Maui, Hawaii)

STORMS ON THE SUN

Light conveys still more information about the properties of its source. We have seen how the type and amount of material present may be deduced from the spectrum. We are, however, also able to determine the motion of gases on the Sun. This is a result of the wave-like nature of light.

We know that the Fraunhofer lines arise because the atoms in the cooler outer layers absorb light at specific, characteristic wavelengths. For the hydrogen-alpha line, 500 000 000 000 000 (5×10^{14}) wave-crests arrive every second. As it passes through the solar atmosphere, this light is greatly weakened, much more strongly than light whose crests arrive with a slightly greater or lesser frequency. This is because hydrogen absorbs light at just one specific wavelength, such as that of the hydrogen-alpha line, but not at neighbouring wavelengths. Such an explanation is not fully accurate, however.

The Sun's surface is not quiet; the bubbles of gas that form the granulation are constantly rising and sinking. The solar material moves like the storm-tossed surface of the sea. Let us look at a specific point in the centre of the disk, where the material is, for the moment, moving directly upwards (and thus towards us). Let us now observe the wave-crests emitted by the rising material at a particular frequency. Let us assume that the waves were originally emitted at the frequency mentioned earlier. At what frequency would we receive them? Because the material emitting the light is moving towards us, each crest has a slightly shorter distance to cover than its predecessor. The arriving crests are separated by a shorter interval of time than when they were emitted. In other words: the distance between the crests is shorter, so the wavelength is reduced. The whole spectrum of the rising cloud of gas on the Sun is slightly shifted to shorter wavelengths, i.e., towards the blue end of the spectrum.

Let us now consider the light from a point on the Sun where, in the course of its inward and outward motion, the material is moving downwards, i.e., away from us. This time every crest has a slightly longer distance to cover than its predecessor, which was emitted when the material was closer to us. The distance between crests is therefore greater when it reaches us than it was when the light was emitted. The spectrum is therefore slightly shifted towards the red. This effect is named after the Austrian physicist, Christian Doppler (1803–1853). The Doppler effect allows astronomers to determine, from just the spectrum, the velocity at which even the most distant stars are approaching us, or receding from us.

The light entering our slit and producing a spectrum comes from all the points of the selected strip of the solar disk. Yet the solar material is in motion, as we have seen with the granulation (Figure 4.2). What effect does this have on the spectrum of the narrow strip of the Sun under examination? Every point along the slit produces its own spectrum, in the form of a small horizontal strip, which may be shifted towards the blue or red end of the spectrum. For instance, points on the slit that receive light from areas of the Sun that at that particular instant are rising, have a spectrum that is slightly shifted towards the blue, and those that are falling one shifted towards the red. The solar spectrum therefore consists of innumerable small adjacent, horizontal, individual spectra, some of which are shifted towards the blue and some towards the red. This may be seen particularly clearly in the dark Fraunhofer lines. They are not straight, but are wavy, as shown in Figure 5.5. As the bubbles of gas rise and fall, kinks occur in the spectral line. These may be either to the right or to the left, or they may disappear completely, or switch from one side to the other. The degree to which the lines are bowed enables solar physicists to determine the velocities of the bubbles of gas. In this way we know that the bubbles forming the granulation move, on average, with a velocity of one kilometre per second. This is 3600 km/h, an utterly ferocious speed by any terrestrial standards!

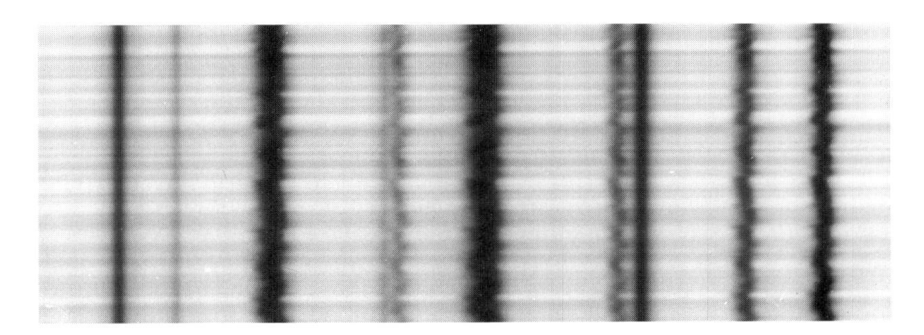

Figure 5.5. A section of the spectrum of a small strip of the solar disk, which captures the granulation. The spectral lines are distorted in many places through the Doppler effect, being bowed either towards the right (the red) or towards the left (the blue): they have a wavy structure. Note that there are a few spectral lines that do not show this effect, such as the two on the left-hand side of the photograph. These are not produced on the Sun, but are caused by atoms in the Earth's atmosphere, which absorb sunlight at certain specific wavelengths. Turbulence on Earth has far smaller velocities, so its effects are not visible in this spectrum

The curves in the Fraunhofer lines may be used to make the motions of the solar surface directly visible. Look again at one of the Fraunhofer lines in Figure 5.5. Figure 5.6 shows a typical line schematically. Imagine that we look at the spectrum through a filter that passes just the narrow band of light in the region bounded by the two thin, parallel lines. The regions of the solar surface where the material is moving upwards and the Fraunhofer line is, because of the Doppler effect, shifted into the pass-band of the filter thus appear dark. When the material is sinking, the Fraunhofer line is shifted in the other direction. The corresponding point of the strip therefore appears bright. Velocities are directly visible in terms of brightness.

Figure 5.6. How velocities on the Sun may be seen directly. Spectral lines are not sharp, black lines in the spectrum; they possess a certain width. Using an interference filter, it is possible to select a narrow strip of the right-hand flank of the line. At points where the line bends towards the right (because of the Doppler effect) we see less light than at points where it bends towards the left. If the solar disk is photographed through this filter, we obtain an image of the Sun where the masses of gas that are moving towards the observer appear bright, while those moving away appear dark. Another method of capturing the motions on the Sun is shown in Figure 6.9

In fact, a Fraunhofer line is not a simple dark line. Moving across a spectral line, the brightness gradually declines, reaches a minimum at the centre, and then increases again when we reach the other edge. This is shown in greater detail in Figure 5.6. If, by using a spectrograph or a Lyot filter, we select just light from the right-hand 'flank' of the line, there is more light where the material is sinking directly away from us and less, where it is rising directly towards us, because the line is shifted to the right, or to the left, respectively. If the whole solar disk is trailed across the slit, then we obtain an image that is bright where material is rising and dark where it is sinking. Such an image is shown in Figure 6.10.

Using the technique just described, the velocities in the solar granulation may be measured. In addition it may be used to obtain the rotational velocities of regions near the poles, at latitudes where there are no sunspots that may be used to determine the rotation directly. Recently, this technique has also led to the discovery, that as well as its rotation and the rising and falling motions in the granulation, the Sun also vibrates with a regularity that had not previously been suspected. We shall discuss this further in Chapter 10.

Apart from revealing the regions of the Sun that appear bright or dark at a specific wavelength, and showing us how the solar material is rising or falling, the spectrum also enables us to determine the Sun's magnetism.

MAGNETIC FIELDS AND SPECTRA

The Nobel Prize for physics for 1902 was awarded to the Dutchman Pieter Zeeman (1865–1943) for a discovery that this previously unknown physicist made in 1894. The question had been in the air for a long time. Michael Faraday (1791–1867) had tested, in 1862, whether light might be influenced by magnetic fields. But he had not been able to determine any such effect. Zeeman, who was inspired by Faraday said: 'If a Faraday saw the possibility ... it would probably be of value to repeat the experiment with the excellent spectroscopic equipment available nowadays, which, to my knowledge, has not been done by anyone else.'

Zeeman investigated the yellow spectral line produced by sodium. Its colour may be seen in nearly all flames. It is responsible for the yellow colour that an otherwise nearly colourless gas flame assumes when cooking salt is introduced into it. This line occurs in the solar spectrum in absorption, and therefore appears dark. Its light is removed from the Sun's radiation. Fraunhofer designated it with the letter D, since when it has been known as the *sodium D-line*. When Zeeman placed a luminous flame in a strong magnetic field, he saw that the yellow spectral line broadened. Very soon Zeeman had refined his experimental technique to such an extent that he was able to see that a strong magnetic field split the line into two narrow, juxtaposed, individual lines. The distance between the individual components increased with increasing strength of the magnetic field. In Chapter 3 we briefly mentioned the polarisation properties of light. We will not discuss this more fully here, but it may be noted that the polarisation of Fraunhofer lines that have been split by a magnetic field provides information about the direction of the magnetic field.

It is not surprising that light, which is emitted or absorbed by atoms, may be affected by a magnetic field. Light is emitted by charged atomic particles (electrons) that are in motion. When electrical charges move, however, magnetic fields exert a force upon them. Magnetic fields therefore affect the electrons responsible for emitting light, and hence they change the radiation emitted by atoms. In the simplest case, two subsidiary lines are formed alongside the original line—regardless of whether the light is in absorption or in emission. If a spectral line is found to be split through the Zeeman effect, it is possible to determine the strength and direction of the field responsible. Figure 5.7 shows schematically how a spectral line is affected by a magnetic field on the Sun.

A few years after Zeeman had shown how a magnetic field alters spectral

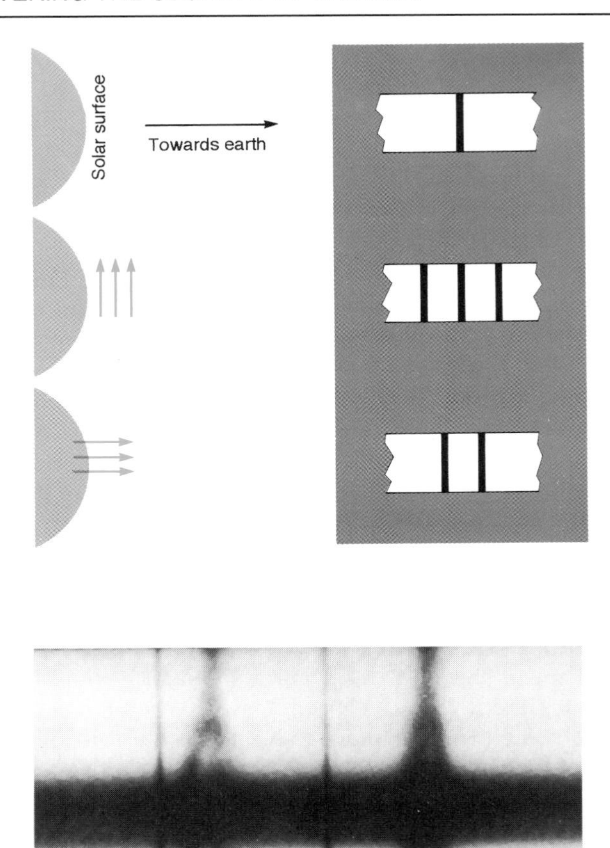

Figure 5.7. Top: When a spectrograph on Earth is pointed towards the solar surface, absorption lines appear in the spectrum (top right). However, magnetic fields above the solar surface alter the appearance of spectral lines in the spectrum. If the field lines run across the line of sight, each line is split into three components (centre right). If, however, they are directed towards or away from the observer, the line is split into two. Bottom: An enlarged view of four lines in the spectrum of a small portion of the solar surface. The dark stripe running across the centre of the picture is the (very faint) spectrum of a sunspot, which is crossed by the lines that are of interest. The brighter spectra above and below this are those of the surrounding surface. Note how the lines near the spot are split by the strong magnetic fields

lines, Hale tried to use this effect to search for magnetic fields on the Sun. Hale was, by that time, recognised as a solar researcher. He had photographed the Sun in hydrogen light and in the light of other elements with his spectroheliograph. His *spectroheliograms,* as they were called, showed the Sun as it had never been seen before.

Hale gained the impression that spectroheliograms in the light of hydrogen-alpha showed a whirlpool-type structure around sunspots, which reminded him of the spiral currents flowing inside a magnetic coil. He therefore assumed that strong magnetic fields were present in sunspots. This led him to discover that the sunspot cycle was not 11 years long.

6

THE MAGNETIC SUN

If the Sun had no magnetic field it would be as boring as many astronomers seem to believe it is.

Robert Leighton

At a total solar eclipse, the corona surrounds the obscured disk of the Sun like a milky-white halo, and its rays often stretch far out into space (Figure 6.1). Their shape resembles that of magnetic field lines that extend from a magnetised body. It had therefore been suspected for a long time that the Sun might be a giant magnet. It could, so to speak, be seen with the naked eye. George Ellery Hale, who developed the spectroheliograph, discovered that the Sun is a very special form of magnet, one that continually changes its properties, but which at the same time, exhibits certain regularities. Even today, however, we still do not fully understand the Sun's magnetism.

MAGNETIC FIELDS IN SUNSPOTS

Hale compared the spectrum of light from sunspots with that of the quiet solar surface, and found that many Fraunhofer lines appeared stronger in the sunspot spectrum. He also compared the solar spectrum with that of luminous gases in the laboratory. The lower the temperature of the gases, the stronger the lines in the laboratory spectra became. From this Hale concluded that sunspots are cooler than the surrounding solar surface. We now know that the gases in sunspots have temperatures of 4000°C or less, whereas their surroundings are at a temperature of 5500°C.

Using the Zeeman effect, Hale also began to search for magnetic fields on the Sun. Let us recap: a Fraunhofer line may split into several components in a magnetic field, as shown in Figure 5.7. In fact, it had been noted many years before that sunspot spectra occasionally showed spectral lines that were split. Hale was the first to succeed in showing that the light from the individual components of these lines exhibited the polarisation predicted by the Zeeman effect. The cause of the splitting was indeed magnetic fields.

Figure 6.1. The total solar eclipse of 7 March 1970. By using a special filter it has been possible to give the correct exposure to the corona, both in regions close to the Sun's limb and where the rays reach far out into space (photo: High Altitude Observatory, Boulder, Colorado)

In 1908, Hale published a paper entitled 'On the Probable Existence of a Magnetic Field in Sun-spots'. He had again compared sunspot spectra with those of luminous gases in the laboratory. This time, however, he placed the laboratory gases in strong magnetic fields, so that the degree of splitting of their spectral lines by the Zeeman effect was the same as that of the sunspot lines. He discovered that the fields were inconceivably strong. At that time the strengths of magnetic fields were measured in gauss. The strength of the Earth's magnetic field is about half a gauss. The quiet solar surface has fields of about the same strength. But the magnetic fields that Hale measured in the sunspots had strengths of 2900 gauss! Hale was also able to determine the direction of the magnetic fields. Many sunspots were 'north poles', and many 'south poles'. What had previously been seen as merely dark spots on the solar disk turned out to be extremely strong magnetic regions.

THE MAGNETIC CYCLE

Sunspots tend to occur in pairs, not at random positions, but oriented East–West. Because the Sun's rotation carries spots from East to West, the western spot is known as the *preceding spot*, and the eastern one, the *following spot*.

When Hale discovered the magnetic fields of sunspots, he noticed that the two spots of a pair were always of different polarity. If one was a north pole, the other was a south pole. In addition, the polarities of sunspots in the northern and southern hemispheres are reversed. If all the preceding spots in the northern hemisphere are north poles, and all the following spots south poles, then all the preceding spots in the southern hemisphere are south poles, and all the following ones north poles.

The greatest surprise, however, came during the following sunspot minimum, around 1913. The last spots of the declining cycle were visible near the equator, when the first spots of the new cycle began to appear at higher latitudes. But in both hemispheres their polarities were opposite to those of the remaining spots of the previous cycle. Figure 6.2 schematically illustrates the sunspot polarity over a few recent solar cycles. We may see from this that in fact, instead of repeating every 11 years, the solar cycle actually lasts 22 years.

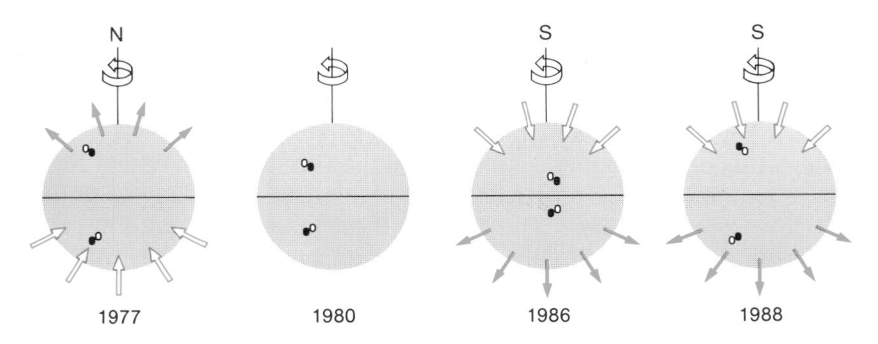

Figure 6.2. Diagram illustrating the Sun's magnetic cycle. In 1977, the Sun's North Pole was also the magnetic North Pole. In the northern hemisphere, the preceding spots were of northern polarity, and following spots were of southern polarity. Around 1980, the polar fields disappeared. The spots in the two hemispheres had the same polarities as three years earlier. In 1986, the cycle was nearing its end and the spots were close to the equator. The North Pole had acquired southern polarity, and the South Pole, northern. In 1988, sunspots of the new cycle were visible at high latitudes. In the northern hemisphere the preceding and following spots now had southern and northern polarity, respectively. In the southern hemisphere the polarities were also reversed

What is the situation now with regard to the assumption that the ray structure of the corona is somehow linked to a magnetic field? Hale had shown that, because of its sunspots, the Sun is saturated all over with small north and south magnetic poles. Does it, like the Earth, also posses a large-scale field structure at its poles? Are the Sun's poles also magnetic poles? Hale was actually able to detect uniform fields at the poles. Their strength amounted to just a few gauss, so the fields at the poles cannot be compared with those found in sunspots. They are, in fact, only slightly stronger than the Earth's magnetic field that governs our compass needles. However, the direction of the Sun's general magnetic field (as it is normally termed), alters with a 22-year period. The Sun's polarity reverses every 11 years. This is also indicated in Figure 6.2.

We have already seen (p. 31) that the Earth's magnetic field is also subject to major changes over the course of thousands of years. There are probably similar mechanisms at work producing the magnetic fields of the Sun and Earth and causing them to alter.

Nowadays, interference filters and magnetographs are used to determine magnetic fields on the Sun. If photographs are taken in the light of a spectral line that has been split into two or three components (as shown in Figure 5.7), suitable polarising filters enable one to obtain images of the Sun where regions of one polarity appear bright, and those of the opposite polarity are dark.

Figure 6.3a shows just such a 'magnetogram'. The Sun's North Pole is at top, and East is to the left. The solar rotation carries the surface from left to right. The picture clearly shows how the preceding spots in the northern hemisphere have one polarity, and the following ones the opposite, while the situation in the southern hemisphere is the reverse.

Closer study of the magnetic polarities on the Sun shows that there are also extensive regions that are only weakly magnetic, but which have a uniform polarity over the whole area. The fields in these magnetic regions are much weaker than those in 'true' sunspots. The regions cannot be distinguished optically from neighbouring regions, and the fields may by determined only by the splitting of the spectral lines.

Figure 6.3b shows the Sun in 1985 at the time of sunspot minimum. Small white and black points reveal tiny areas of different magnetic polarity, which cannot be recognised in visible light. It will be seen that the Sun's North Pole (at top in the illustration) appears predominantly dark, and the South Pole, mainly light. This indicates that the poles do not have a uniform magnetic field that is approximately the same strength everywhere, but that the field crosses the surface in small bundles which appear white or black in the picture. The magnetic field at the poles that Hale discovered therefore consists of a large number of individual bundles of field lines. As the solar cycle progresses, one of the Sun's polar regions becomes covered in white points, and the other in black. Eleven years later, everything reverses. The pole that had white regions now turns black, and vice versa.

The Sun's magnetic fields are not completely invisible. It is possible to see that above the areas where the magnetic field lines pass through the solar surface, the overlying layers are particularly bright in the light of the calcium K line (Figure 6.4).

Let us just sum up the course of events on the Sun over the 22-year cycle.

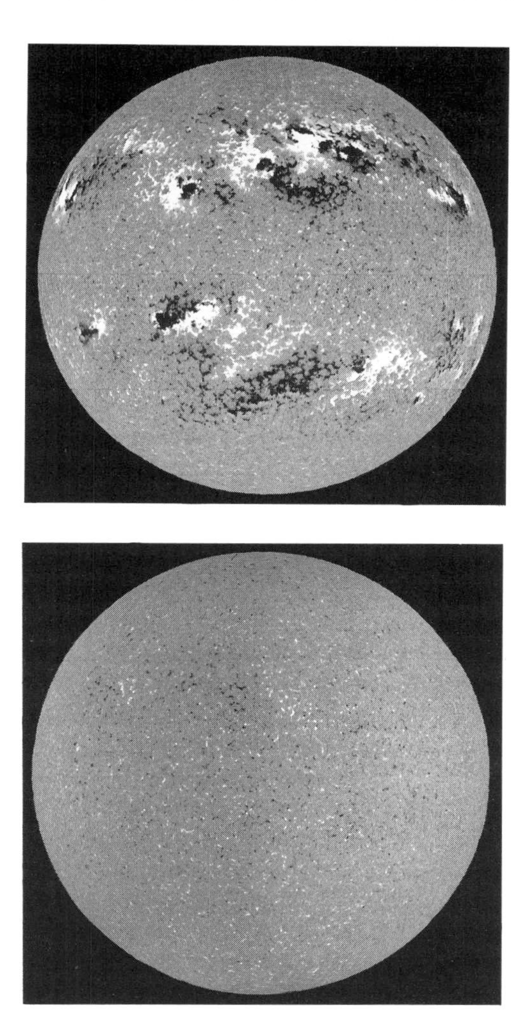

Figure 6.3. (a) Magnetogram of the Sun for 12 February 1989. North is at top, East to the left. Regions of northern polarity appear white, and those of southern, black. The Sun's rotation is carrying the spots from left to right of the picture. In the northern hemisphere, the preceding and following spots are of southern and northern polarities, respectively. In the southern hemisphere, the opposite applies (photo: National Solar Observatory, Tucson, Arizona). (b) Magnetogram of the Sun for 27 December 1985, during sunspot minimum. There are no large magnetic regions, but just small, nearly point-like spots of both polarities. At the Sun's North Pole there are more black points, and more white ones at the South Pole. This indicates that the polar fields actually consist of small, individual, bundles of magnetic field lines (photo: National Solar Observatory, Tucson, Arizona)

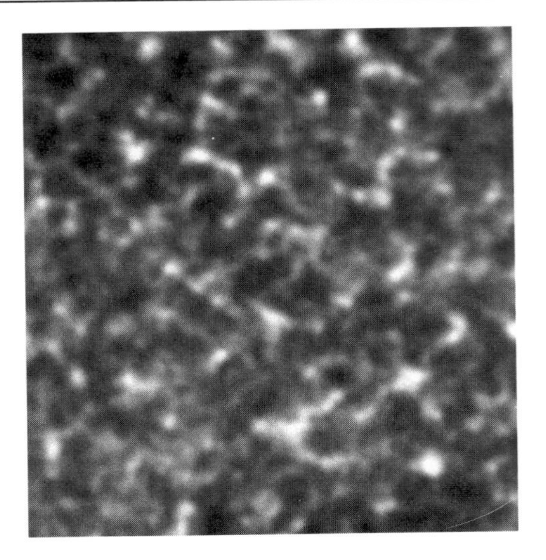

Figure 6.4. A portion of the solar surface, taken in the light of the violet calcium line, shows the Sun's 'calcium network' as a pattern of bright lines (photo: H. Wöhl, Kiepenheuer Institute for Solar Physics, Freiburg)

At the beginning of a cycle, pairs of spots appear in both hemispheres. They appear in two zones around latitudes 30° North and South. In the northern hemisphere, the preceding and following spots have, respectively, northern and southern magnetic polarity. In the southern hemisphere, on the other hand, the spot with southern polarity is the preceding one, and the one with northern, the following spot. Over the course of the next few years, spots become more numerous. As they appear and disappear, they gradually get closer to the equator, just as described by Spörer and Carrington. About 10 years after the beginning of the cycle, the spots close to the equator decline in numbers. We have reached sunspot minimum. The first spots of the new cycle appear, again in two zones at 30° North and South. But now the magnetic polarity is reversed. In the pairs of spots in the northern hemisphere, the preceding spot has southern polarity, and the following one, northern polarity, while in the southern hemisphere, the spot with northern polarity is ahead of its twin. Over the next 11 years the sequence of events is the same as before. In other words, the two zones in which spots occur slowly move towards the equator, although the magnetic fields in the spots have the opposite polarity to those shown 11 years before. When sunspot minimum arrives and the first spots of the new cycle appear, they have the same magnetic polarity as the spots in the penultimate cycle. After 22 years we are back where we started.

Occasionally the last spots of the preceding cycle have still to disappear when the first spots at high latitude usher in the new cycle. They may be recognised by their reversed magnetic polarity. The new cycle appears 'impatient' and the two cycles 'overlap'. For a short time, pairs of spots may be observed in the same hemisphere, where some of the preceding spots are of one polarity, and others are of the opposite polarity. After the last spots of the old cycle have disappeared, order prevails again on the Sun. At least that is what was thought until comparatively recently. This orderly picture was disturbed in 1953.

EPHEMERAL SUNSPOTS

On 13 August 1953, an astronomer, Clifford Bennett, was observing the Sun at the McMath-Hulbert Observatory of the University of Michigan. The last sunspot minimum had occurred some eight years previously. The next would be in about three years. The two sunspot zones were at latitudes of about 15° North and South. In a few years the first spots of the next cycle would appear at higher latitudes. At 12:45 Universal Time, Bennett discovered a tiny spot in the northern hemisphere at a position where there should not be a spot. We have already seen that sunspots hardly ever occur at higher latitudes than about 40° (Figure 2.9). But this newly formed spot was at 52° North! It was immediately examined with the spectroheliograph in both calcium and hydrogen light. There were bright lines at the spot's position.

At Mount Wilson Observatory in California, the Sun's magnetic fields were regularly monitored. On 13 August, at about 17:30, a relatively restricted area of northern magnetic polarity was discovered at the position of the spot found at Michigan. Alerted by their colleagues in Michigan, the California researchers investigated this unusual area in greater detail, and discovered that a pair of spots was present. The spot that had been discovered initially was the preceding spot. There was a second, much fainter, following spot with southern magnetic polarity—in the middle of sunspot cycle, when the rest of the northern hemisphere had preceding and following spots with, respectively, southern and northern polarities! This new pair of high-latitude spots contradicted the established law of polarity. It seemed as if a forerunner of the new cycle, which was not expected for some years, had not only appeared in the wrong place, but also at the wrong time. On the Butterfly Diagram shown in Figure 2.9, this extraordinary spot is marked with a cross.

While the magnetic properties of this remarkable pair of spots were being investigated in California, they had long since disappeared in visible light. As early as 14:00 Universal Time, and thus only slightly more than an hour after the discovery, the astronomers in Michigan had lost sight of the spot. The magnetic fields lasted somewhat longer. They finally disappeared about 23:00.

When the solar researcher Helen W. Dodson of the McMath-Hulbert Observatory published a short paper about this remarkable spot, she described the 'ephemeral' nature of the phenomenon. This was the birth of a new concept in solar physics. 'Ephemeral' means fleeting or short-lived. 'Ephemerides' are the tables in which astronomers list daily positions of celestial bodies. It is also the scientific name for the mayfly. We nowadays speak of *ephemeral active regions* when we mean these short-lived sunspots. It is not surprising that their existence was not even suspected for such a long time, because most of them are invisible. They may only be recognised on magnetograms of the Sun. The spot of 13 August 1953 was by no means typical of ephemeral active regions, but they came to be called after it.

Once attention had been drawn to the phenomenon, careful searches were carried out for short-lived magnetic regions. The result was a surprise. Every day perhaps as many as 100 pairs of such spots appear and disappear. They are invisible in normal light and detectable only in magnetograms or in spectroheliograph images. They appear suddenly, are generally fully developed after a day, and disappear a day later. Like sunspots, they are particularly numerous at sunspot maximum. They are distributed over the whole of the solar surface, however, and appear at high latitudes where sunspots are never found.

To make this easier to understand, imagine that the Sun's latitude and longitude grid, with its equator and poles, has been transferred to the surface of the Earth. Sunspots occur in two zones, which stretch to about 32° on either side of the equator. On Earth that corresponds to a region from the equator to Algeria in the north, and to South Africa in the south. The short-lived spot of 13 August 1953, however, appeared at a latitude corresponding to that of Berlin or Saskatoon in Canada. Ephemeral active regions may occur as high as the latitude of Norway or the outer regions of Antarctica.

This is illustrated in Figure 6.5. The ephemeral active regions extend the 'wings' in the Butterfly Diagram to such an extent that they overlap. During sunspot maximum, spots occur in the equatorial regions: say, in the Congo and Angola on our model. At the same time, there are ephemeral active regions at the latitudes of Central Europe and the Falkland Islands. As the years go by, not only do the sunspot zones move towards the equator, but so do the ephemeral active regions. When the latter reach latitudes of about 30°, sunspots appear there. Many solar physicists believe that there is thus a continuous transition between the zones in which the ephemeral active regions occur and those with sunspots. This is not to say that ephemeral regions do not appear and disappear at lower latitudes. On the contrary, they are also seen between spots and spot-groups. The ephemeral active regions seem to appear at high latitudes as forerunners of the new cycle, while the old one is still in full swing near the equator. This is also indicated by their magnetic polarity. High-latitude ephemeral regions show the same magnetic polarity as the next cycle.

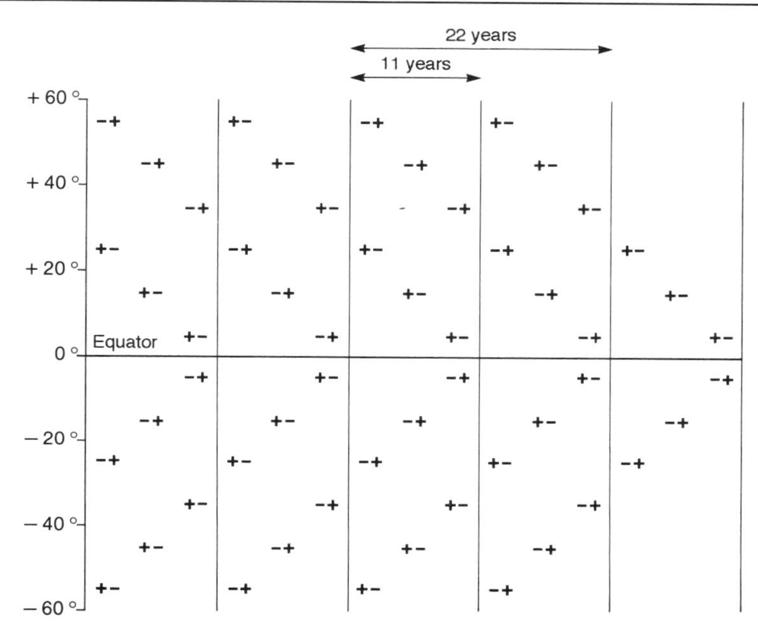

Figure 6.5. A schematic representation of the overlapping magnetic cycles on the Sun. Until comparatively recently only the magnetic phenomena associated with sunspots were known, and these were confined to bands reaching from the equator to latitudes of about 30° North and South (Figure 6.2). Since the discovery of ephemeral active regions, these bands must be extended to about 60° North and South. In doing so it becomes obvious that while the spots belonging to one cycle are appearing at latitudes 30° North and South, the ephemeral regions at high latitudes are actually forerunners of the next cycle

The study of these ephemeral active regions turns out to be extremely difficult, because motions within the granulation repeatedly bring regions of differing magnetic polarity close together. This creates small regions that suddenly appear to combine, and thus, for a very short time, become a sort of miniature sunspot pair. True ephemeral active regions consist of a pair of regions of opposite polarity that remain linked together throughout their (albeit short) lifetime. The others are merely brought together accidentally, and are not mutually linked in the same way. True and false pairs are difficult to distinguish from one another. Ephemeral regions do not strictly adhere to the rules governing the polarity of the preceding and following spots, unlike sunspots themselves, which are strictly governed by them. When taken statistically, however, the ephemeral regions that appear at high latitudes do seem to anticipate the polarity of the forthcoming cycle, long before the first true sunspot appears after sunspot minimum.

The ephemeral regions suggest that the Sun's magnetic behaviour consists not of two 11-year cycles with different polarity, but rather of two 22-year cycles of opposite polarities, which follow one another at 11-year intervals, and thus partially overlap. At any one time, two such cycles are observed on the Sun. By the time the older one exhibits sunspots, the newer one is already responsible for ephemeral regions at high latitudes.

THE SOLAR PROMINENCES

We have already discussed the red tongues of flame that may be seen projecting into the darkened sky at a solar eclipse (the prominences) in Chapter 4. These masses of gas appear to float freely above the Sun's surface (Figure 4.5). We have also seen that they may be recognised when seen against the bright solar disk. Figure 5.4 shows numerous dark, worm-like objects stretching across the surface. They are clouds of hydrogen gas that absorb the H-alpha light from the radiation arising in the Sun interior. When they are seen from 'above' (so to speak) it is obvious that they are actually flat, upright 'sheets' of cooler gas. They are known as *filaments*. When seen from the side at the solar limb, they are found to tower to heights of some 50000 km. They are surrounded by the solar corona, which has a temperature of about two million degrees, but they are significantly cooler, with temperatures of a few thousand degrees. They are therefore 200–300 times as dense as the thin coronal gas. The mass contained within a small filament is comparable with that of a small terrestrial mountain. They are long-lived objects. They frequently persist over several solar rotations, disappearing over the western limb, and reappearing over the eastern limb about two weeks later. They obviously have some connection with the Sun's magnetic field. They do not occur at random across the disk, but are found at the boundary between magnetic fields of opposite polarity. Why do they not fall down onto the solar surface? We shall see in Chapter 8 that they are held in suspension by the magnetic fields.

Frequently they simply disappear, and at other times they 'explode' and shoot upwards at high velocities. They are often flung out so fast by unseen forces that their material escapes from the Sun, and may even reach as far as the Earth.

Where does their material come from? It seems to condense from their surroundings, that is, from the solar corona. If coronal gas condenses into filaments, many of which are lost to space, then the corona itself would become exhausted, if it were not supplied with material all the time. What serves to provide a continuous source of material for the corona?

SPICULES

If the Sun's limb is observed in the red hydrogen line, it is by no means flat. Small, luminous spikes may be seen projecting into the dark sky, like grass in

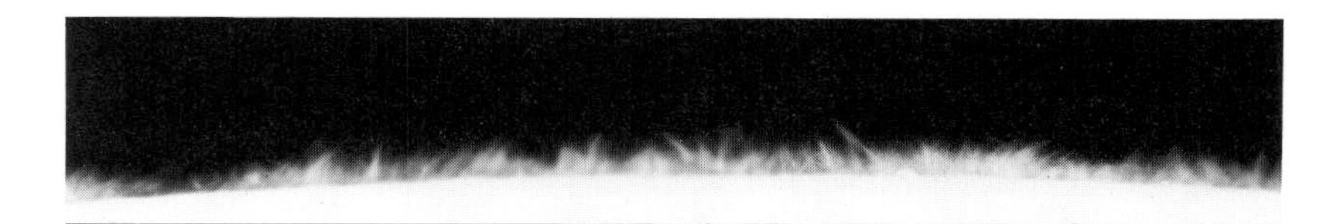

Figure 6.6. Spicules seen at the solar limb in the light of the red hydrogen line (photo: R.B. Dunn, Sacramento Peak Observatory)

a meadow (Figure 6.6). The Italian solar researcher, Angelo Secchi, gave them the name of *spicules*. The individual 'blades of grass' shoot up to heights of 15 000 km at speeds of about 20 km/s. Their width amounts to about 2000 km. They disappear after about 10 minutes, giving way to new 'blades of grass'. As the old spicules fade, new ones continuously arise. Only rising material is observed within them. Nothing appears to fall back onto the Sun. Is material continuously shot out of the surface of the Sun into the corona?

Spicules may be observed not only at the solar limb, but also (through suitable filters) on the disk, when it is found that the 'grass' does not grow the same all over the surface, but occurs in 'tufts' or miniature 'hedgerows' (Figure 6.7).

Figure 6.7. Viewed in the light of the red hydrogen line, spicules on the solar disk appear to cluster in 'hedgerows' (photo: R.B. Dunn, Sacramento Peak Observatory)

LUMINOUS CALCIUM ON THE SUN

The appearance of the Sun in red hydrogen light is different from that in white light. We are actually looking at a layer that is about 1000 km above the solar surface. The image that we obtain if we observe the Sun in the violet calcium line, which Fraunhofer called the K line, is completely different again. Here we are looking at yet higher layers.

In calcium light, sunspots appear bright. In general, calcium-light images (Figure 6.8) show bright regions, where magnetograms show strong magnetic fields. This is why the regions of sunspots appear bright. But other features are visible as well. The whole of the Sun, even areas where there are no sunspots, is covered by a network of bright lines. This *calcium network* has already been seen in Figure 6.4. The bright lines coincide with the hedgerows of spicules. In addition, improved magnetic-field measurements show that weak magnetic fields project out of the Sun at the same points.

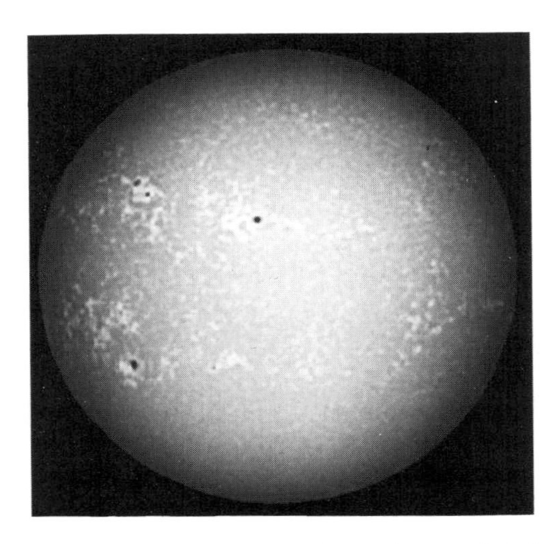

Figure 6.8. The Sun in the light of the calcium line. There are bright spots where there are magnetic fields

Why are these regions bright in calcium light? There is, in fact, a simple explanation. The K-line of calcium is a sensitive thermometer, which gives us the temperature of the layers that are 1000 km above the surface. The areas that are bright in calcium light are slightly hotter.

Why should it be hotter, where the magnetic field is stronger, than elsewhere? And why are the spicules found just there? What is responsible for the individual cells of the calcium network? The answers to all these questions came in 1959.

GRANULES AND SUPERGRANULES

The American physicist, Robert Leighton, who teaches at the California Institute of Technology in Pasadena, is also a master of highly refined observational techniques. Together with his collaborators Robert Noyes and George Simon, he photographed the Sun in the light of several spectral lines. They obtained two simultaneous images of the Sun by using two slits, which they set to the two wings of the lines they were investigating.

Because of the Doppler effect, the red slit receives more light from portions of the Sun that are approaching us, and the blue slit more from regions that are moving away (Figure 6.9). The image recorded by the red slit therefore shows regions of the Sun that are moving towards us as brighter, and the blue image shows receding regions as brighter. If a transparent negative is made of the latter image and superimposed on a transparent positive of the former, we obtain an image that is bright where material is approaching us, and dark where it is receding. Naturally, this applies to the material in the layer from which the light originates.

Leighton's images showed a distinct structure. Such an image is shown in

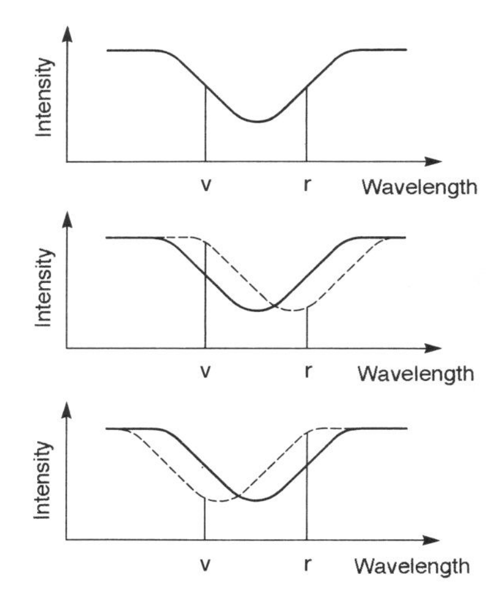

Figure 6.9. Top: The profile of a Fraunhofer line is symmetrical. If, by using Lyot filters, light is obtained from narrow bands on the left-hand, or violet side (v) of the spectrum, and also from the right-hand, longer-wavelength or red side (r), both images are therefore equal in brightness. Centre: If material is moving away from us, however, because of the Doppler effect, the line is shifted towards the right (broken curve). The light at (v) is significantly stronger than that at (r). Bottom: If the material being observed is moving towards us, the image at the wavelength (v) is fainter. By using a suitable combination of images and their negatives, it is possible to obtain an image of the Sun, where material that is moving towards us appears bright, and material moving away appears dark. Such an image is shown in Figure 6.10

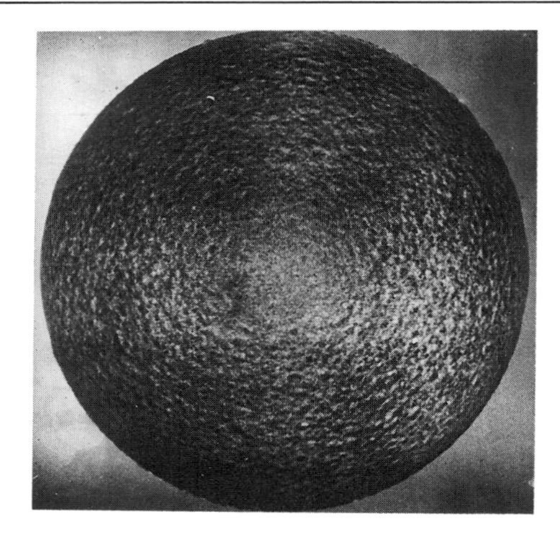

Figure 6.10. A 'Doppler image' of the Sun, obtained by the means described in Figure 6.9. The regions of the solar surface that are moving towards us appear bright, and those moving away are dark. The structure that is visible is the supergranulation. Away from the centre of the disk, the cells are being observed at an angle, and consist of bright and dark areas. The bright portions are closer to the centre of the disk, and material is therefore moving towards us. The portions closer to the limb are dark, because material is moving away. The reason is because material is rising in the centre of the supergranulation cells and spreading out towards the edges (photo: California Institute of Technology)

Figure 6.10. The cells are particularly distinct in the zones that are neither too close to the limb nor too close to the centre of the disk. The side of each cell that is closest to the centre of the disk appears bright. The side closest to the solar limb, on the other hand, appears dark. This is to be expected if material is flowing out from the centre of each cell towards the edge.

The cells discovered by Leighton's team have no connection with the granulation. Granules are visible in white light, and may be seen using amateur-sized telescopes. The granular appearance of the solar surface caused by the granulation arises from strong temperature differences in the deeper layers of the solar atmosphere. The newly discovered *supergranulation* is not visible in white light. It occurs in layers in which the temperature is nearly uniform. The granules, which have been known for a long time, are small, generally about 1000 km in diameter. The supergranules are about three times the size. Whereas the small granules form and disperse within minutes, supergranules last, on average, about a day, before they are replaced by further supergranules. In the granulation cells, the gas rises and falls with velocities of a few kilometres per second. As we have already seen, the motion within the supergranules is mainly horizontal and occurs at velocities of about 500 m/s. This is why the supergranules themselves cannot be detected in the monochromatic light of a single spectral line at the centre of the solar disk (see Figure 6.10), because flows of material on the Sun are only detectable (through the Doppler effect) out towards the Sun's limb. Only there does gas that is streaming horizontally across the solar surface move towards or away from us.

We find that the pattern formed by the supergranulation cells coincides with the calcium network. From this we also know that the magnetic network is somehow involved.

We have seen that processes on the Sun, such as prominences and the calcium network (and the supergranulation that is associated with it), are always related to the Sun's magnetic field. This connection is even more noticeable in the case of the eruptions that are occasionally observed to take place on the Sun's surface.

ERUPTIONS ON THE SUN

We have already met Richard Carrington, who discovered the law governing solar rotation and the Butterfly Diagram. At 11:20 on 1 September 1859, he was projecting the image of the Sun onto a screen held some distance behind the eyepiece. He had just drawn a spot group, when two neighbouring regions in the centre of the image suddenly brightened. The bright spots visibly increased in size. Although he was an experienced observer, he had never seen anything like this, so he immediately tried to find someone to witness it. When he came back a minute later, nothing could be seen. The event had lasted no more than five minutes.

During the next night, workers in observatories that were monitoring the Earth's magnetic field measured major disturbances with unprecedented strengths, in what is known as a *magnetic storm*. Carrington suspected that the event that he had seen on the Sun was responsible. He was, however, cautious. He wrote that merely because two events happen to occur at apparently the same time, it is not possible to conclude that they actually related. He added, 'One swallow does not make a summer.'

Since then, these explosive brightenings on the Sun, which are known as *flares*, have been observed frequently. Generally they do not last more than about an hour (Figure 6.11). Carrington was right; they do cause magnetic disturbances on Earth. Auroral events are also closely linked to them.

Flares always occur where strong magnetic fields are observed. They appear to be concentrations of the magnetic fields in their neighbourhood. Often a flare causes a previously quiescent filament to 'erupt', and to be expelled out into the corona, or even farther into space, as if by some unseen hand.

We have already described the corona, which becomes visible during a solar eclipse, when the Moon covers the disk of the Sun, and the region immediately surrounding the Sun is no longer swamped by light from the disk. Ever since

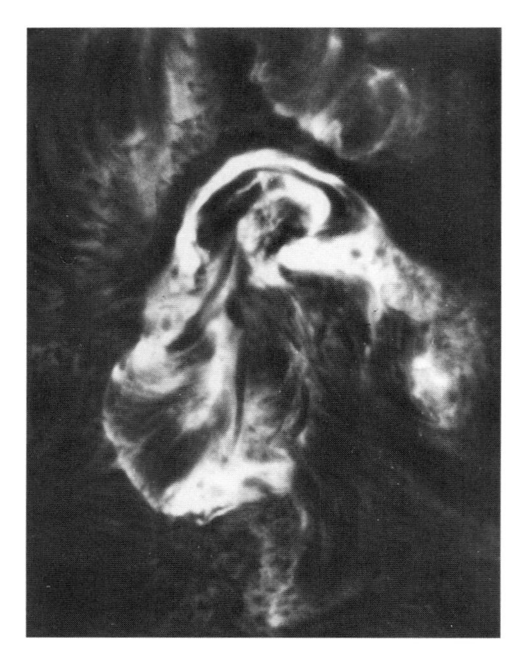

Figure 6.11. A flare, photographed in red hydrogen-alpha light on 3 July 1974 (photo: A. Bruzek, Kiepenheuer Institute for Solar Physics, Freiburg)

it was realised that coronium, the element, does not exist (see p. 64), but that iron atoms that had lost half of their electrons were responsible for producing the green coronal line, we have known that the corona is hot. Only then do the atoms move so fast that they strip electrons from one another when they collide. Even when it proved possible to produce reasonable numbers of such highly ionised atoms here on Earth the green coronal line was still not observed, because it only appears when density of the iron vapour is as low as it is in the Sun's corona. In fact, the density of matter there is lower than the best vacuum obtainable on Earth.

The low-density material in the solar corona is extremely hot. The temperatures are estimated to be around two million degrees. When compared with the temperature of the corona, the solar surface, at just 5500 °C, is ice-cold. The Sun attains such extreme temperatures only deep in its interior. At a temperature of one million degrees, material emits radiation at X-ray wavelengths. The X-ray radiation from the Sun helps us to determine more about the corona.

At first sight, our theories of heat and mechanics do not seem to be valid on the Sun. At sunspot maximum the surface is dominated by dark magnetic spots and, above them, leaf-like condensations that appear to move around freely. Above them yet again extends the corona, which is far hotter than the surface beneath it. Brilliant flares suddenly erupt, and invisible forces hurl prominences outwards so violently that they do not fall back to the surface. Why does solar material behave in such remarkable ways? The answer is that it is in a special state. Solar material is in the form of a *plasma*.

7

SOLAR PLASMA

> From our school-days we are used to dividing all materials into three characteristic forms—solid, liquid, and gaseous. In recent years, however, the fourth state, which, because of its distinctive properties has been called plasma, has come to be the centre of ever greater interest.
>
> D.A. Frank-Kamenezki

We are accustomed to materials, whether they are solid, liquid, or gaseous, obeying the law of gravity. They drop down towards the ground if they are not supported. Their motions are governed by the laws of mechanics. This seems normal to us. Yet only a tiny fraction of the matter in the universe is in this simple state. By far the largest portion is found in stars, where it is by no means 'normal': it is hot and possesses unusual properties. It is still affected by gravity, but in general other forces are stronger.

Whereas, under normal conditions, no electrical current can flow through air, the gas in the flame of a candle is electrically conducting. The gas within stars also carries electrical currents. This is because at the high temperatures that prevail within stars, atoms are easily 'broken apart'. Some of the electrons that orbit the atomic nuclei are ejected, and they then move freely among the remaining portions of the atoms (known as *ions*) which are now positively charged. This is why the gas is said to be *ionised*. We have already described what are iron ions in connection with the extremely high temperatures in the solar corona. Very similar properties are found in metals. Although, in general, the ions in a metal are not able to change their positions relative to one another, the electrons are able to move freely among them. This is why metals are good electrical conductors. For the same reasons, the gas in a flame, and the gas within a star, are electrically conducting, because both are ionised.

However, it is not just the high temperature of a gas that makes it electrically conducting. Gases in space may also be conducting at lower temperatures. In the nearly empty space between the stars, the weak starlight is nevertheless sufficiently strong to remove at least one electron from most of the heavier elements. These electrons are sufficient to cause the gas between the stars (known as the *interstellar gas*), to be electrically conducting. The gas in the Sun is in the same state. The electrical conductivity of solar gas is about the same as that of copper on Earth.

The copper in an electrical cable is a solid, even though electrons flow through it. The ions hold the metal together strongly, and force has to be expended to bend the cable. The ionised gas within stars can flow. The interstellar gas moves freely throughout the vast spaces between the stars. The nearest that anything on Earth comes to the state in which cosmic gases occur is found in mercury. As a metal it is electrically conducting, and thus possesses free electrons. But its ions are also free to move relative to one another, because at room temperature mercury is liquid. Matter that contains both electrons and ions that are free to move has been called a *plasma* ever since the term was suggested by the American physicist Irving Langmuir (1881–1957) in 1928. Liquid mercury is a plasma, but so is the gas in a flame, the gas inside the Sun, and the gas between the stars.

The Soviet physicist Frank-Kamenezki wrote in 1963: 'Physicists first described plasma fairly recently, but they had seen it much earlier. Plasma is the principal architect of the impressive phenomena of lightning and of the aurora. Anyone who has ever had the 'pleasure' of causing a short-circuit, has been dealing with a plasma. The sparks that jump from one conductor to another consist of a plasma caused by an electrical discharge in the air. When we walk through a city at night and look at illuminated advertisements, we do not realize that the tubes contain a glowing plasma of the noble gases neon or argon. Any material that is heated to a sufficiently high temperature turns into a plasma. . . . An ordinary flame has a certain degree of electrical conductivity; it is—even if only very slightly—ionized, and thus a plasma.'

Nearly all the material in the universe is ionised. Electrical and, above all, magnetic, forces sometimes control the motion of a plasma more strongly than does gravity. The freedom of the electrons and ions to move relative to one another causes the plasma to have completely different properties to those of material in the forms that are more familiar to us. To understand the properties of matter in the form of plasma, we first need to consider a few points relating to magnetism.

MAGNETIC FIELDS

We know that magnets attract pieces of iron and steel. In his lecture for beginners, a well-known German experimental physicist was regularly in the habit of hauling his assistants, who kept firm hold of an iron plate, into the air, using a magnet attached to a block and tackle. His colleagues said that he used to choose his assistants mainly by their weight, to ensure that this impressive demonstration worked.

Magnetic fields have always impressed people, because they appear to act through empty space and can, for example, move a piece of iron, without the magnet touching it. We know that the needle of a compass is governed by the invisible magnetic field of the Earth. We say that the force is transmitted through empty space by a *magnetic field*.

An idea has been devised that makes this easier to understand, and which we shall use extensively in what follows. This is the concept of *magnetic field lines*. They are an indispensable aid to anyone who wants to study the way in which a magnet exerts its force, or the field produced by a current. We shall make use of this concept in many of the following chapters, and it will help us to understand the magnetic phenomena on the Sun. Apart from its advantages, there is just one disadvantage: it does not exist. These 'lines' are imaginary, despite the fact that they may be demonstrated by the very simplest of experiments.

Remember what our physics teacher once showed us? He covered a magnet with a piece of paper. He then scattered iron filings over the flat sheet of paper. As he did so, and particularly if the paper were shaken slightly, the iron filings lined up, and a regular structure appeared. The filings appeared to lie along curved lines that spread out from the area around one of the poles of the magnet in more of less tightly curved arcs that then converged on the other pole. These imaginary lines that appear to be indicated by the iron filings, are known as magnetic field lines. Their appearance around a horseshoe magnet is shown in Figure 7.1.

I have just said that magnetic field lines do not exist, and I must explain this. Close to a magnet, space—whether full of air, or completely empty—is in a

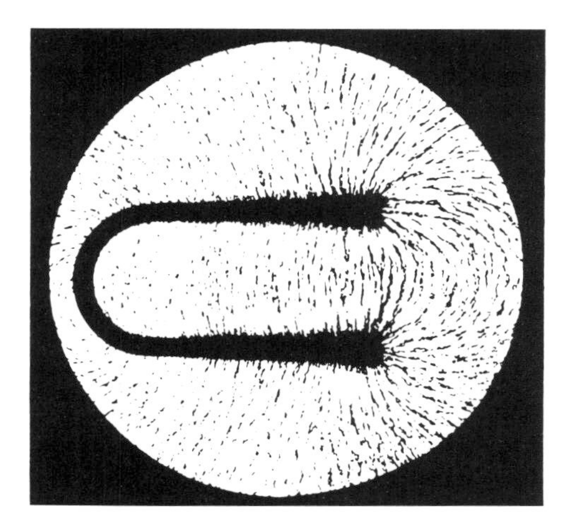

Figure 7.1. Iron filings scattered on a glass plate enable the magnetic field lines around a horseshoe magnet to be seen

special state. Within it, forces are exerted on other magnets, such as compass needles. Scattered iron filings do not fall where they might be expected to under just the force of gravity. They are subject to an invisible force. Physicists are only able to describe the state of space around magnets—i.e., within what is known as a magnetic field—by means of complicated mathematical equations. These may be used to calculate the forces that pull iron filings aside. The equations also describe the direction in which a compass needle affected by a magnet will lie. Physicists can describe all this formally, based on the lectures on electrodynamics and potential theory that they attended when they were students. In general, physical equations are not accessible to the layman. This is just where magnetic field lines help.

The complicated effects of magnetic fields (often difficult to predict) may be described by a few rules. These rules, which we shall come to individually, act as if there were a field consisting of invisible lines, the field lines, whose path could be revealed by iron filings. Frequently the magnetic forces behave as if the (imaginary) field lines are like stretched rubber bands.

Now let us discuss the type of magnetic fields that may occur in plasma, and how they affect the plasma. As mentioned in the introduction, Herr Meyer's dreams will help here.

HERR MEYER IN PLASMALAND

Herr Meyer had spent a hard day by the time he got back to Munich on Saturday evening. Since early that morning he had been walking round the Max Planck Institute for Plasma Physics out at Garching, going from building to building, and from exhibition hall to exhibition hall. It had been the Institute's Open Day, and, like thousands of other visitors, Herr Meyer, who had previously thought of plasma as something to do with medicine, had learnt what a complicated thing a physical plasma is. The physicists had produced it in various clever pieces of experimental apparatus. But no sooner had it been produced than it slipped through their fingers again. Concepts such as ions, electrons, and magnetic fields were all jumbled up in his head as he tried to go to sleep.

Suddenly he found himself in Hamburg. He had just come out of the Dammtor station, when he saw Mr Tompkins on the other side of the road. Highly delighted, he crossed the extremely busy road and went up to his friend. It had been a very long time since they had last seen one another in a different town, and Mr Tompkins was beaming with pleasure.

'Mr Tompkins', said Herr Meyer, 'what are you doing in Germany?'

'I wanted to see the circus here in Hamburg', he replied, pointing to the tents and booths that had been set up on the Moorweide, in the centre of Hamburg.

'The Plasmaland Circus is world-renowned, and I wanted to see it once for myself', he said. 'You know how I like curiosities.'

Herr Meyer knew that it was always interesting to be in Mr Tompkins' company. It would doubtless prove to be the same with the circus, so he said 'If it's alright with you, I should like to come as well.'

It did not take them long to make their way to the ticket office. 'You must be my guest', said Mr Tompkins, and soon came back with two tickets, two sets of cardboard spectacles, and two programmes.

'The spectacles enable you to see magnetic fields', he explained. The view through the spectacles was disappointing. As Herr Meyer looked through the sheets of plastic that substituted for glass, everything appeared exactly the same as it did without spectacles. Nothing seemed bigger or smaller. All the colours were the same as previously.

'Look', said Mr Tompkins, pulling a small piece of metal that had the well-known horseshoe shape of a magnet out of his pocket. Through the spectacles Herr Meyer could now see red lines, which curved from one end of the horseshoe to the other (Figure 7.2).

'It will be more interesting,' said Mr Tompkins, as he leafed through the programme, 'when they switch on the plasma.' When Herr Meyer looked at him enquiringly, Mr Tompkins continued. 'That is the most important thing in the Plasmaland Circus', he explained. 'The equipment here would have brought millions to its inventor, if he had not decided to restrict himself to exhibiting his patent in this circus. When the equipment is switched on, it emits waves that make everything around the machine—air, water, and every object—electrically conducting. Everything suddenly turns into plasma. Anyway, we can see the first act now.'

With these words, he led Herr Meyer into the nearest tent. But before we follow them, we need to consider the question of where magnetic fields are to be found in nature.

Figure 7.2. Mr Tompkins shows Herr Meyer the field lines produced by a horseshoe magnet

ELECTRICAL CURRENTS PRODUCE MAGNETIC FIELDS

Let us imagine a copper wire connecting the two poles of a battery. If an electrical current is flowing through it, the wire may become hot, may glow, or may even melt. Electrical currents and magnetic fields are intimately linked. Figure 7.3a shows the field lines surrounding a wire passing at right angles through the page, as they would be revealed by iron filings. A particularly strong magnetic field is produced by a wire that is wound into a coil. Figures 7.3b and 7.3c show the magnetic fields produced by a single loop of wire, and by the multiple windings found in a coil. Finally, Figure 7.3d shows how iron filings would delineate the magnetic field of a closed coil.

It would seem as though there are two types of magnetism: one appertaining to magnets and the other to currents. But the difference is only apparent. Currents also flow within a magnet. They are responsible for the magnetic field of a horseshoe magnet, for example. All magnetic fields are produced by currents. Even the magnetic field that affects our compass needles is created by currents within the Earth's interior.

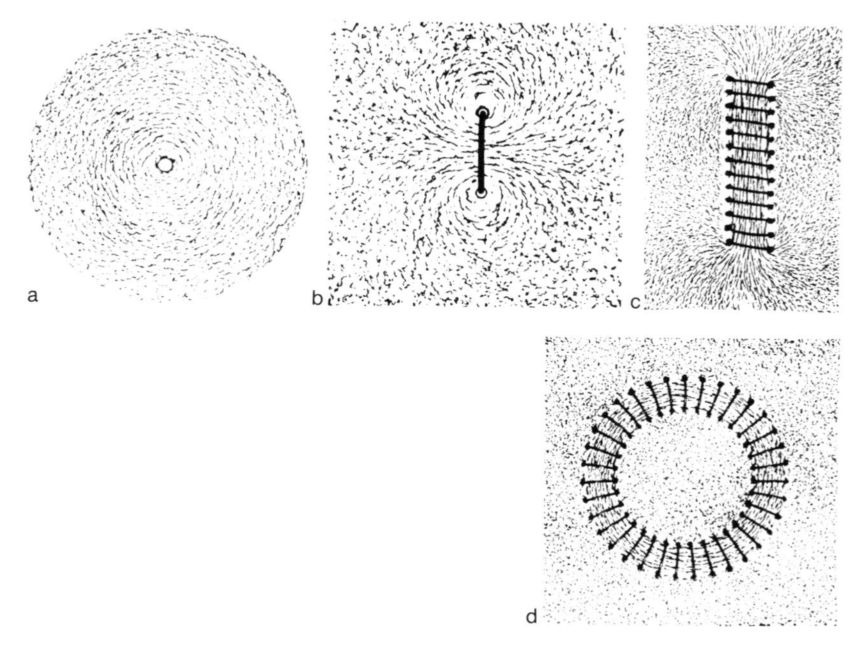

Figure 7.3. The field lines around various conductors carrying current, as revealed by iron filings. (a) A current passing at right-angles to the plane of the page produces circular field lines. (b) The field produced by a single turn of wire, which is at right-angles to the plane of the page. (c) The field obtained with a coil of wire. (d) The field produced by a circular (toroidal) coil of wire

MAGNETIC FIELDS PRODUCE ELECTRICAL CURRENTS

Electric currents produce magnetic fields, but the opposite also applies: magnetic fields produce electrical currents. If an electrically conducting wire is moved in a suitable magnetic field, an electrical current flows within it. An example is a bicycle dynamo. When the dynamo is turned by the wheel, a coil of wire rotates in the field produced by a magnet, creating a current that powers the bicycle's lights.

When we consider the two, fundamentally linked laws that govern the mutual interaction between current and field, it becomes obvious how complicated the conditions within a plasma may be. Imagine that we have a magnetic field in a plasma, which we want to set in motion. The material is either a fluid or a gas, so there is no difficulty in allowing it to flow. But note that plasma is a conductor. So we are moving a conductor relative to the magnetic field. This means that within the conductor, i.e., in the plasma, a current is created. The latter produces a new magnetic field, in addition to the one that already exists. If the plasma moves, it affects the magnetic field.

At first it appears a hopeless task to predict how the magnetic field will be altered. There is, however, a simple rule that allows us to predict what changes will occur in a magnetic field within a plasma—even when we are dealing with the most involved magnetic fields or highly complicated currents. This is where the concept of magnetic field lines will help. First we need to simplify the problem, which we may do by assuming that the plasma is an infinitely good conductor. We know that any wire exerts a certain resistance to an electrical current. This is because the electrons that are moving in the metal wire are not completely free to move. When they pass the metal ions, they occasionally collide, so they do not move without any hindrance. In poor conductors the friction between the ions and the electrons is high, whereas in good ones it is low. So we first need to assume that our plasma is an exceptionally good conductor, and that there is no interaction between the electrons and ions. Then things become simple. This is what Herr Meyer saw when he visited the first tent.

HERR MEYER, THE CLOWN AND THE DOG

Mr Tompkins had obtained two seats in the front row. From his seat, Herr Meyer was able to see the whole ring clearly.

'Everything here looks just as it does in any ordinary circus, with sawdust in the ring. What we cannot see are the big magnetic coils that are hidden below the floor and up above the ceiling', explained Mr Tompkins. Indeed, so far Herr Meyer had seen nothing unusual, apart from the large plate that blocked their view upwards, and which was the same size as the ground within the ring.

'A magnetic field will shortly be switched on between the two plates. Exactly three minutes later the plasma generators will be switched on. Perhaps we should put on our spectacles.'

They perched the spectacles on their noses. Everything still looked exactly the same to Herr Meyer as it did before he put on the spectacles. He asked Mr Tompkins for the magnet and held his hand in between the two ends. The lines passed straight through his hand and came out the other side. Suddenly he felt a strong jerk on the magnet, which jumped out of his hand and fell on the ground.

'The magnetic field has been switched on!' cried Mr Tompkins. Herr Meyer looked at the ring. Between the ground and the top plate lines were visible, which ran straight up and down, like taut threads. Only at the edge of the ring were the lines crooked, and there they bowed out into the area where the spectators were sitting. They included the first two rows of seats and, on looking more closely, Herr Meyer could see that field lines were visible around him.

It must have been the magnetic field that wrenched the horseshoe magnet out of his hand. He bent down and picked it up. The magnet still showed the lines coming out of one end and curving over to the other, but at some distance from the magnet they were distorted by the magnetic field in the ring. Just as Herr Meyer noticed that the new field lines also passed through his hand, there was a burst of applause and a clown and his dog entered the ring. The field lines still ran straight up and down. The clown and the dog moved freely through them. This did not surprise Herr Meyer, because he had gathered from his hand that human bodies did not interact with magnetic fields, whether that caused by a horseshoe magnet or the one now prevailing within the ring.

Suddenly, the whole of the tent was bathed in red lighting. There was a gasp from the crowd. Mr Tompkins explained: 'Now everything inside here has become plasma. You will see the difference right away.'

Herr Meyer sudden realised that the magnetic field lines appeared to be attached to his hand. Lines passing through his hand remained in the same place, even when he moved his hand. They followed his hand and dragged other field lines aside as they did so. When Herr Meyer brought the small magnet, which he was still holding in his right hand, near the palm of his left, he observed that the magnetic field lines that were outside his hand when the red lighting came on, now refused to pass through it. Instead they bent away and clustered densely together in front of his palm to avoid entering it. They seemed to be frightened of it. Only field lines that passed through his hand before the plasma generators were switched on still passed through it.

Suddenly his attention was drawn by what was happening in the ring. Music began to play. What happened next is shown in Figure 7.4. The clown tried to catch the dog, who was running round the circle, but he was unable to get near it. But it was not the runners that captured Herr Meyer's attention,

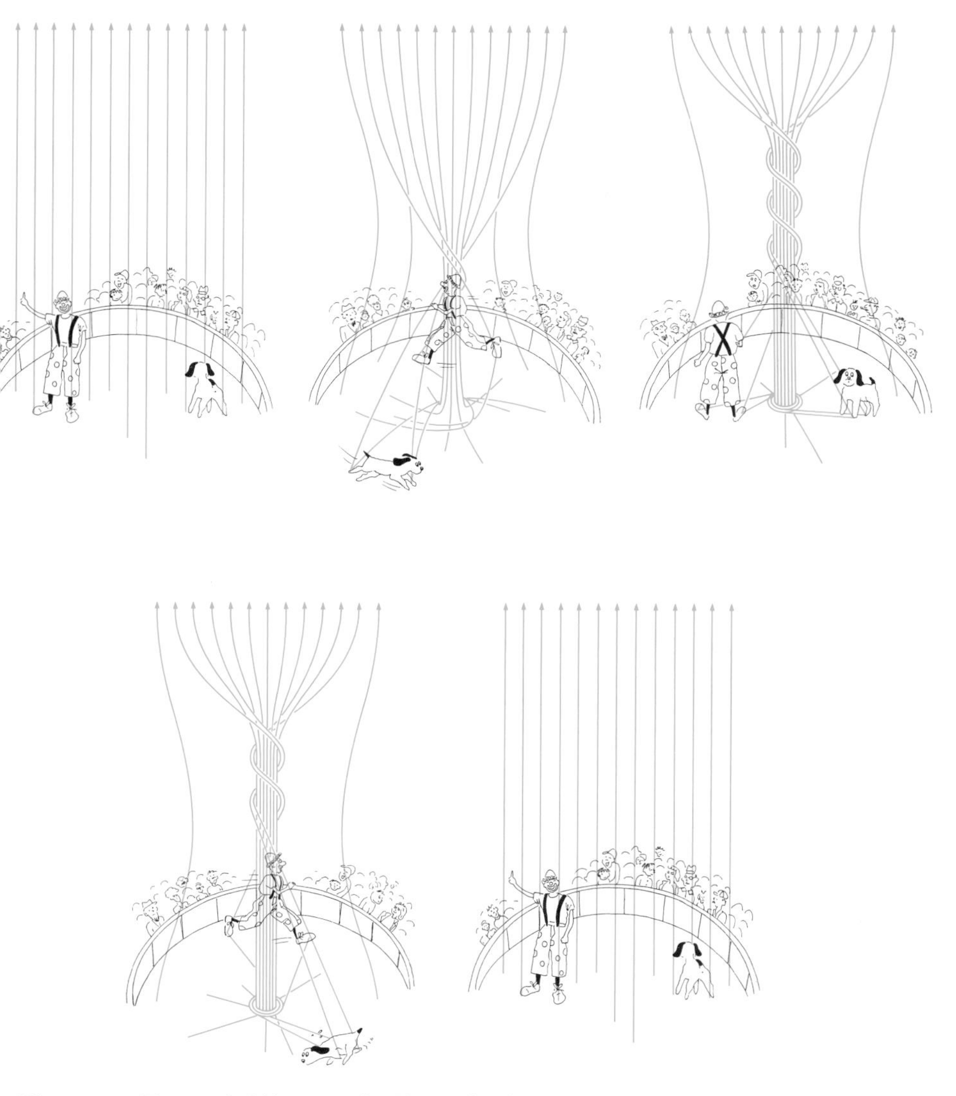

Figure 7.4. Frozen field lines in the Plasmaland Circus. At the beginning of the act the field lines ran perpendicularly up and down inside the ring. Because the clown and dog were already there, when they ran round, they dragged the magnetic field lines round behind them. So the lines became twisted up, both in the air and at ground level. When the performers reversed their direction, the field lines became untwisted until, at the end, they again ran straight up and down

but the magnetic field lines. These lines, which had previously passed through their bodies, now seemed to resemble lengths of string that followed their movements. They were compressed together tighter and tighter at both top and bottom as the clown and the dog continued to run round without stopping. Eventually, the lines were so closely packed together that Herr Meyer could hardly make out individual lines in the densest parts. At a signal from the music, the two stopped, turned round, and ran round the ring in the opposite direction. Again the magnetic field lines followed them. Because the direction was now reversed, they became untangled, just as if one were unravelling a twisted piece of rope. Just as the field lines became straight again, the dog stopped dead. The clown also came to a halt. Now the field lines were more or less straight again, just as they were originally.

The plasma state was switched off. The red lighting disappeared, and the field lines vanished. Plasmaland had become an ordinary circus again. The spectators clapped, the clown bowed, and the dog stood up on its hind legs and showed that it approved of the applause.

'We have seen an example of frozen field lines' said Mr Tompkins. 'Whilst everything, including the clown and the dog, was plasma, the magnetic field lines moved with matter. They were "frozen" into the performers' bodies.' Now Herr Meyer finally understood what the plasma physicists had meant when they spoke about frozen field lines during the Open Day. 'If they had had a clown and a dog', he thought pensively, 'everyone would have understood.'

FROZEN MAGNETIC FIELD LINES

What occurred in the circus tent with the clown, the dog, and the plasma created in the surrounding air was very complicated. Let us begin with a simpler example, where the initial magnetic field lines run straight up and down. This time our plasma does not contain either a clown or a dog. It is simply an evenly conducting gas. Let us now move a layer to one side, as is shown by the horizontal arrows in Figure 7.5. We have seen that this will cause currents to flow, which will alter the magnetic field. What does this look like? It is simple: the field lines moved with the material, just as if they had been moved along with it. The field lines are fixed within the plasma! This is why they are known as *frozen field lines.*

Another feature of the interaction between plasma and magnetic fields is closely associated with this. We can explain it by another thought experiment. Between the poles of a large magnet there is a magnetic field, which may again be represented by parallel field lines. Outside the field there is a cloud of electrically conducting plasma (of infinitely low resistance) that is essentially free from any magnetic field. Now let us move the plasma towards the magnetic field. Figure 7.6 (left), shows what happens.

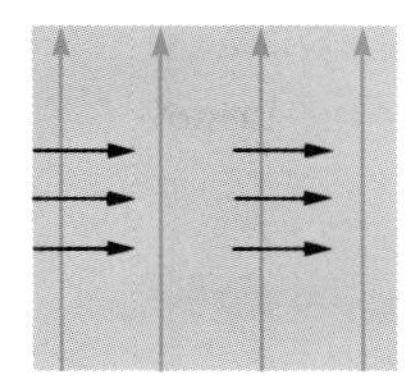

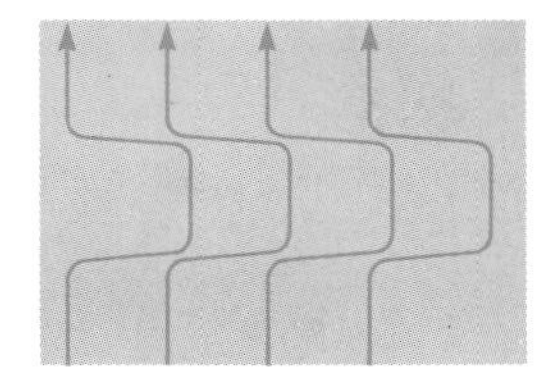

Figure 7.5. Parallel field lines within a plasma (left), the centre of which suddenly moves towards the right (black arrows). The frozen field lines are also shifted as a result of the motion (right)

The magnetic field lines avoid the plasma! To express this more accurately: When the electrically conducting plasma is moved towards a magnetic field, the latter induces currents within it, which themselves produce a magnetic field, which, together with the pre-existing field causes the magnetic field lines to curve as shown in the figure. This may also be expressed more informally: *If there was no magnetic field in the plasma originally, then none is able to penetrate it later.* The opposite also applies (Figure 7.6 right). If initially there is a magnetic field within the plasma—however induced—it is unable to escape from the plasma. If, for example, the magnetic field becomes weaker, currents immediately flow within the plasma, that tend to maintain the field. With a plasma that is an infinitely good conductor, there is nothing that will cause the currents to wane. To express it in an even simpler form: *If a plasma contains a magnetic field to begin with, there is no way of cancelling it later.*

Does it make any sense to speak of a magnetic field within a plasma when it is impossible to introduce one from outside? It is not a contradiction. We can, for example, imagine ordinary air, which is not conducting, and therefore possesses no currents that can resist the introduction of a magnetic field. Once it contains a field, it is possible to heat the air so much that the atoms become ionised. The electrons that escape are free to move and convert the gas into a plasma that is traversed by a magnetic field. If we make the further assumption that the electrical conductivity is infinite, then the magnetic field cannot escape while the material remains in the plasma state.

So far, we have imagined that the plasma moves, and have seen how the magnetic field reacts, which introduced the concept of frozen magnetic field lines. If the plasma is stationary, the field lines that are present remain unchanged. If the plasma moves, then the field lines follow the plasma. But if there was no magnetic field in the plasma originally, then none may be induced within it.

Our plasma with an infinite conductivity is an idealisation that we have used to become familiar with its most important properties. Despite this, some of the properties of a plasma apply—at least to a certain extent–even to some of the electrical conductors that we encounter in ordinary life.

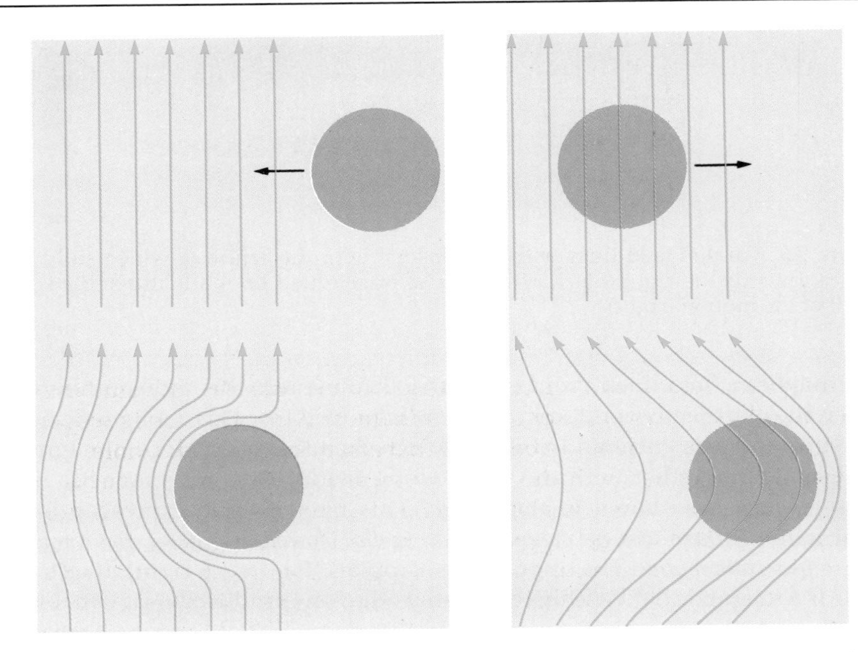

Figure 7.6. The appearance of frozen field lines. Left:If we attempt to introduce a ball of plasma into a magnetic field, the lines bend away and avoid the plasma. Right: If we try to remove a plasma from a magnetic field, then the field lines follow the motion, because the lines remain frozen into the plasma

If we try to insert a copper bar, 10 centimetres thick, into a strong magnetic field, the magnetic field lines do not pass into its interior immediately. Electrical currents are produced within the copper, and their magnetic fields tend to oppose those of the external magnetic field. It takes about a second for the opposing currents to die away and for the magnetic field to enter the copper. The same sort of thing occurs it we suddenly pull the bar out of the magnetic field. Again currents are induced that attempt to trap the magnetic field within the bar. So if we try to pull the bar out, the field lines move with it and remain inside the bar for about one second. When the bar is removed from the magnetic field it carries the field with it for a second, so the magnetic field remains frozen for that time. Then the currents sustaining it decay, and the magnetic field disappears.

This effect becomes greater, the larger the bar of copper. If we want a rough estimate of the time, we can use a simple rule of thumb: twice the thickness, four times the duration; three times the thickness, nine times the duration of the magnetic field. Imagine an experiment with a bar one metre thick: the magnetic field would then take 100 seconds to become established within it, and the same amount of time to decay. That is nearly two minutes. The larger the piece of copper, the longer the induction and decay times. Imagine a piece of copper the size of the Earth. The corresponding times are already more than 100 million years. We can see that the rules governing frozen magnetic field lines are important even with ordinary conductivities, when we begin to approach cosmic dimensions.

THE FROZEN MAGNETIC FIELDS IN SUNSPOTS

The average electrical conductivity of the material in the Sun is approximately the same as that of our bar of copper. For a diameter of 1.4 million kilometres, the decay time for the Sun's magnetic field is more than 4.6 thousand million years, which is our estimate of its age. Whatever types of changes we see on the Sun, they cannot possibly be fields that affect the Sun as a whole.

Even with sunspots, which extend over a relatively small area, we can deduce something about the decay times for magnetic fields in a plasma from our rule of thumb. A large sunspot may perhaps extend for about 30 000 kilometres. Even with approximate figures for the electrical conductivity, we find that the magnetic fields in a sunspot can disappear in a period of a few years at most. But sunspots appear and disappear in shorter periods. We therefore must assume that the magnetic field of a sunspot exists long before the spot becomes visible, and that it continues to exist for a similar period after the spot disappears. A sunspot must be merely transitory evidence of a much longer-lasting, but invisible, feature.

If the magnetic fields of sunspots live longer than the sunspots themselves, where are they beforehand, and where do they go afterwards? We have seen that sunspots generally occur in pairs of opposite magnetic polarity. This gave solar researchers the idea that for most of the time the magnetic fields are hidden below the solar surface in the form of long, horizontal strands. Let us assume that there is a magnetic field below the surface, whose field lines are not evenly distributed within the plasma, but are instead twisted into a sort of magnetic 'rope' (Figure 7.7). If a loop of the rope reaches the surface, together, of course, with its plasma, then the field lines can break through the surface and extend out into the overlying atmosphere. A portion of the material can flow back along the field lines. The magnetic field itself extends far out into the atmosphere. It displaces the atmospheric plasma and replaces it with its own plasma, which is derived from deeper layers. Where the loop leaves the solar surface, and also where it re-enters the solar interior, we see two spots of opposite polarity, because at one point the field is directed towards us and at the other away from us. We have already seen that sunspots tend to occur in pairs with opposite magnetic polarity. This strongly agrees with the concept of a magnetic rope.

But why do sunspots appear dark? The solution lies in a form of interaction between the magnetic field and plasma that we have not yet considered.

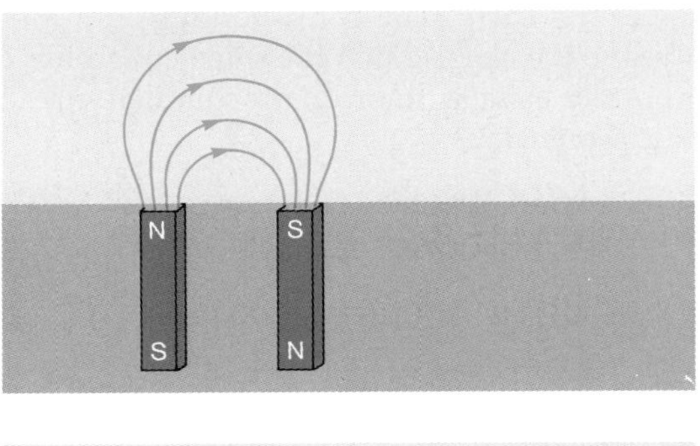

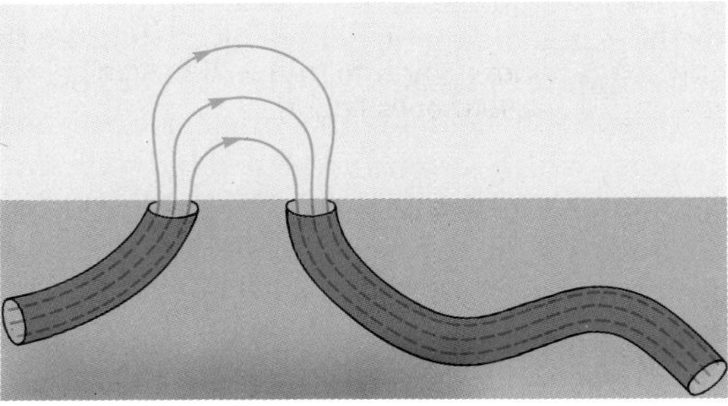

Figure 7.7. How we must imagine the magnetic field in a pair of sunspots. Top: The field lines that would be produced by a pair of magnets beneath the solar surface. From outside we see one spot with northern polarity and another with southern. Bottom: If the magnetic field forms a magnetic 'rope' beneath the solar surface that breaks through the surface at a certain point, then the magnetic field created in the atmosphere is similar to the previous case

THE MAGNETIC-FIELD STRENGTH

Let us return to the plasma in Figure 7.6, left, which we are attempting to introduce into a magnetic field. As we introduce the plasma, we feel a resistance, just as if the magnetic field were repelling the foreign material. The reason for this is that magnetic fields exert a force. It was because of the forces produced that magnetism was originally discovered.

When we think of the great variety of magnetic fields that can exist, it would seem nearly impossible to guess in which direction the magnetic field within a plasma may exert a force. But, once again, there is a simple way of considering the problem.

Magnetic field lines should be imagined as being like stretched rubber bands. This explains, for example, why opposite magnetic poles attract (Figure 7.8, left). But that is not all. Magnetic field lines do not just want to contract, they also mutually repel one another. This explains why, for example, similar poles repel one another (Figure 7.8, right). Let us return to

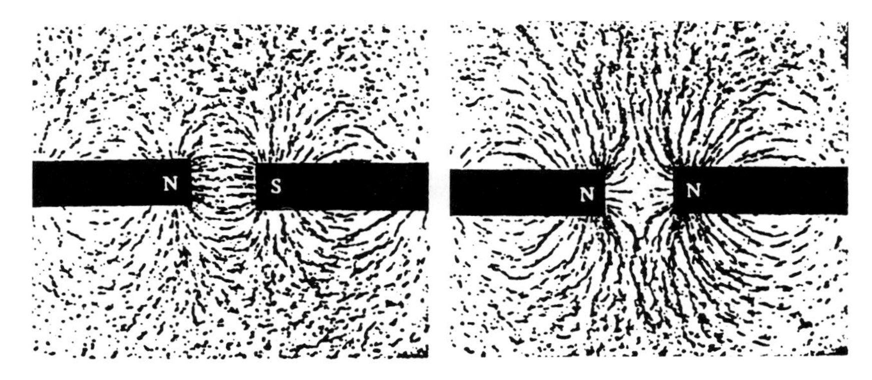

Figure 7.8. The field lines, as shown by iron filings, between two opposite magnetic poles (left), and between two similar ones (right)

the case of a plasma being forced into a magnetic field. Figure 7.6, bottom left, shows the paths of the field lines after the sphere of plasma has been brought close to them. The field now consists of the original (linear) field and an additional field that has been produced by the currents created as the plasma approached. We know approximately what it looks like from the rule governing frozen field lines. Because the field lines are unable to penetrate the plasma, they bend away around the unwelcome invader. To the left of the plasma the field lines are closely bunched together. On the one hand they are attempting to shorten and, on the other, to repel one another. Both effects exert a force acting towards the right. The field tries to repel the intruder. Using the concept of stretched rubber bands that tend to shorten, but also repel one another, we can predict the effects of a magnetic field on a plasma with confidence. These concepts will help us to understand Herr Meyer's next experience.

HERR MEYER AND THE TWO CARDONAS

Mr Tompkins and Herr Meyer let themselves go with the tide of people leaving the first tent, and soon found themselves inside a second, which was even larger. Several zebras were leaving the ring, and they could hear a voice over the loudspeakers: 'And now, ladies and gentlemen, we come to the highlight of our show today: the Two Cardonas. First, Carmen Cardona will carry out her death-defying leap into the ring.'

Herr Meyer looked up and saw a young artiste standing on a small platform at the top of the tent. He had often seen an artist perform this sort of act. There had always been a tank full of water on the ground. The art consisted of jumping properly into the tank in such a way that the water absorbed the full force of the dive. But this time Herr Meyer could not see any sort of tank.

Anyone who jumped from up there must fall straight down onto the hard ground inside the ring, and break their neck.

Up above, the young woman seemed to be getting ready. Suddenly everything went deadly quiet. A faint drum-roll increased the tension. Herr Meyer became worried.

'Can nobody see that there is no tank of water in the ring?' he asked Mr Tompkins.

'Perhaps you ought to put your glasses on', he replied. With this, Herr Meyer realised that he had taken off his cardboard glasses. Looking through them now, he still saw that there was no tank of water, but he realized that up above there were two large coils, to which he had not previously paid any attention. They were producing a strong magnetic field, the lines of which ran horizontally through the performer's body. When the red lighting came on, Herr Meyer knew that everything had turned into plasma.

The woman jumped, and a gasp went through the audience. Herr Meyer saw her fall and noticed that at the same time she dragged the magnetic field lines with her (Figure 7.9). The lines trailed behind her line of fall like a double set of cords. In just a fraction of a second she would crash into the ground. But her fall became slower and slower, and about two metres above the ring, she came to a stop. Immediately—as if attached to rubber cords—she began to rise faster and faster, until she reached the height of the platform, which she cleverly caught hold of. The long magnetic field lines had risen together with the artiste, and once again ran horizontally between the two coils.

There was a burst of applause. The red lighting disappeared, and shortly afterwards so did the magnetic field lines. 'And now,' said the loudspeaker, 'Manuel Cardona will undertake yet another type of death-defying leap.'

Circus hands brought two vertical stands, like advertisement hoardings, into the ring and set them up parallel to one another. The magnetic field was then switched on. This time Herr Meyer still had his glasses on, and so he could immediately and clearly see the straight lines that projected out of one of the stands and went straight across the intervening space and into the other.

A man climbed up to the top of a tall pole, but at the top there was no magnetic field to be seen. Again the plasma generators were turned on. The red lighting and a gentle roll on the drums caused the audience to fall silent (Figure 7.10).

The man jumped, and fell with his arms and body stretched out horizontally. Unlike his sister, he did not pull any magnetic field lines along with him. He had jumped so that he fell directly in between the two uprights on the ground. When he reached the region that was crossed by the magnetic field lines, the latter bunched up beneath his body, and at the same time his fall slowed down. He came to a stop, when the field lines beneath his body were closely pressed together. Then his motion was reversed and he was thrown, as if by a spring, high up into the air, while the field lines resumed

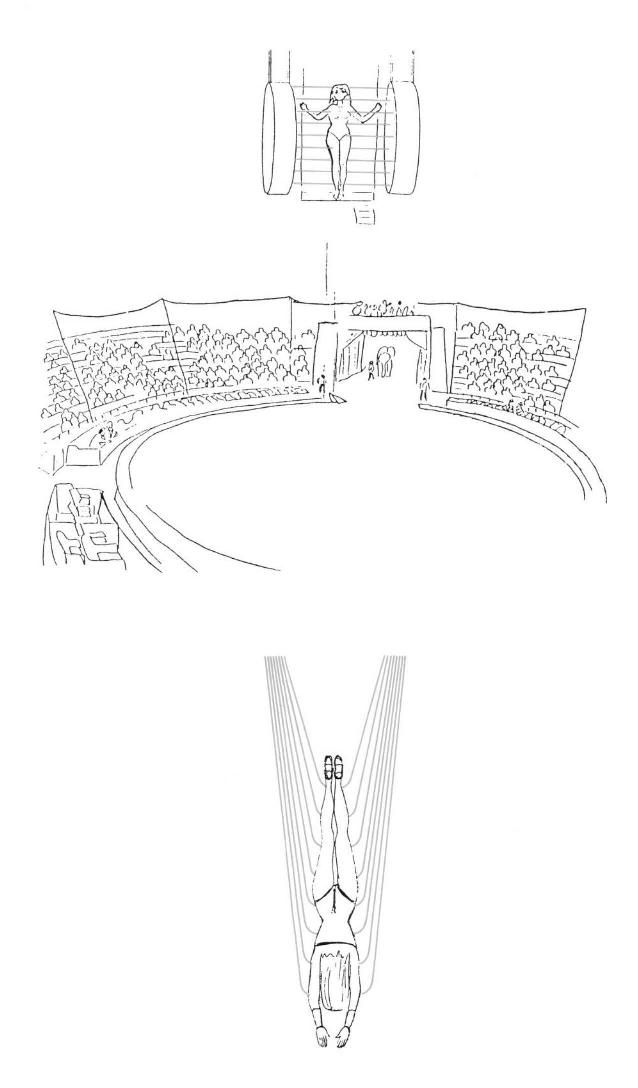

Figure 7.9. Top: In the Plasmaland Circus, the artiste stood on a trapeze between two powerful magnetic coils. When she leaped downward, she carried the magnetic field lines that ran through the plasma of her body with her in her fall (below). The field lines, which tend to contract like elastic cords, halt her fall

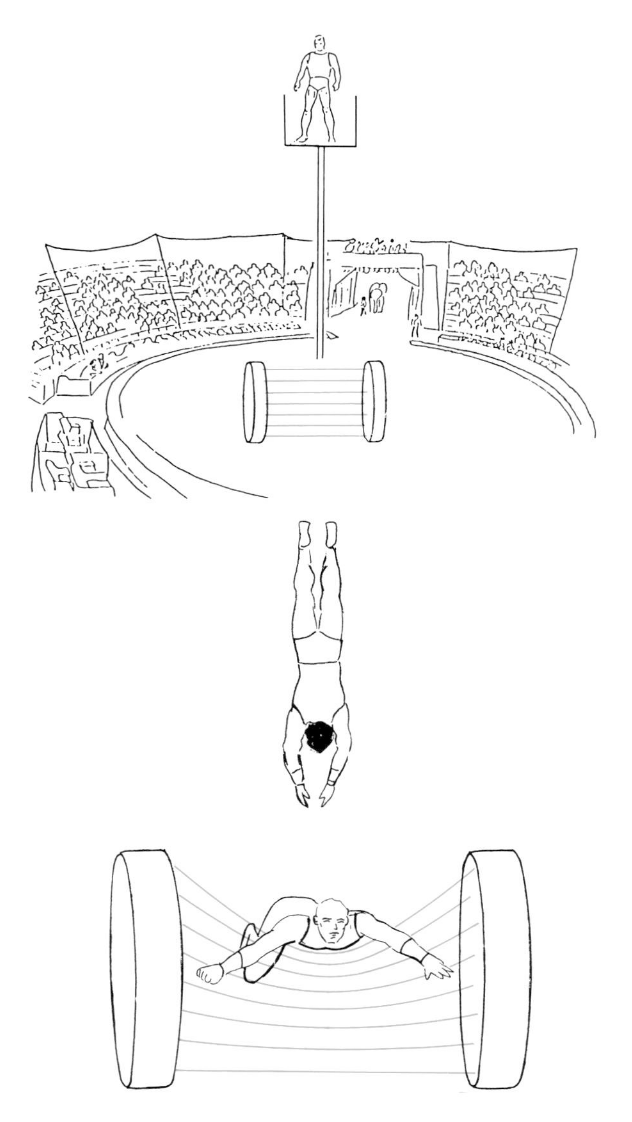

Figure 7.10. Top: The Plasmaland Circus performer jumped from a great height onto a magnetic field that had been produced between two magnetic coils standing in the ring. Centre: Unlike the woman, he did not drag any magnetic field lines behind him. Initially his fall is not braked. Bottom: When he reached the region between the two magnetic coils, the field lines distorted, because they were unable to enter the plasma of his body. The forces thus produced braked his fall and bounced the artist back up into the air

their straight appearance. The acrobat flew upwards, fell back down again, and hit the field lines a second time. As he did so, he twisted his body, so that he fell on his back. Just as if he were using a trampoline, he performed several other twists and turns, to the applause of the crowd. Each time he was flipped back to a lesser and lesser height, until eventually he dropped from a very low height, and fell feet-first onto his magnetic mattress.

The field lines curved aside slightly, and he fell, so to speak, between the lines, which avoided him, and came to rest with his feet on the ground. Then both the plasma and the magnetic field were switched off, and the display by the Two Cardonas was over.

'Two fine examples of the forces exerted by magnetic fields', said Mr Tompkins as they were going out. 'When the woman jumped the field lines were extended. They halted her fall just like elastic cords. In her partner's case the horizontal field lines were compressed together. The forces acted as if the lines were repelling one another. That braked the man's fall, and threw him back into the air.'

'Mr Tompkins is a very interesting man,' thought Herr Meyer, 'but sometimes he acts a bit too much the schoolmaster.'

MAGNETIC FIELDS IN SUNSPOTS

We have seen how frozen magnetic field lines follow the material in which they are trapped, and that therefore the matter within a plasma determines how the magnetic fields may alter. But it is also clear that through the forces that they exert, magnetic fields may determine how matter may move. Which then is predominant in a plasma, the matter or the magnetic field?

In principle, we can imagine two cases. If the magnetic field is weak, its forces are also weak. The material component moves independently of the magnetic forces, and the magnetic field is unable to do anything other than follow the material. But if the magnetic field is strong, the magnetic forces determine what happens, and the material is forced to follow the field. Both cases occur in nature.

The case where the magnetic field determines what happens appears to be found in sunspots. Over vast areas of the solar surface, the energy produced in the interior is transported towards the outside by convection—a boiling motion of the outer layers. (We shall discuss this process, which is responsible for the granulation—see p. 56—in Chapter 9.) In sunspots, however, the magnetic fields are so strong that they determine how material moves in the region. The shapes of the field lines are simply described by the rules that govern elastic bands: they are as short, and as far apart, as possible. Any alteration in their direction requires extremely powerful forces. The convection of the material is thus suppressed in their vicinity. The amount of energy brought to the surface per square metre is therefore less. The temperature of the spots is thus less, which is why sunspots appear darker.

Magnetic fields are not always sufficiently strong to prevent or lessen mass motion. We have seen that there are large magnetic regions on the Sun that, like the ephemeral regions, are undetectable in visible light. In both cases, the magnetic fields are so weak that they are carried along by the motion of the material.

We can now understand why the magnetic fields are so strong around the edges of the supergranulation cells. The situation is shown schematically in Figure 7.11. The supergranulation's horizontal flow of material carries the weak magnetic fields along with it. The magnetic field lines become concentrated at the edges of the cells, where material disappears into the interior. This creates the magnetic network. Now the main question becomes one of why the calcium network coincides with the magnetic network. The reason for this is again because of a special property of a plasma, which we may first explain by another of Herr Meyer's experiences.

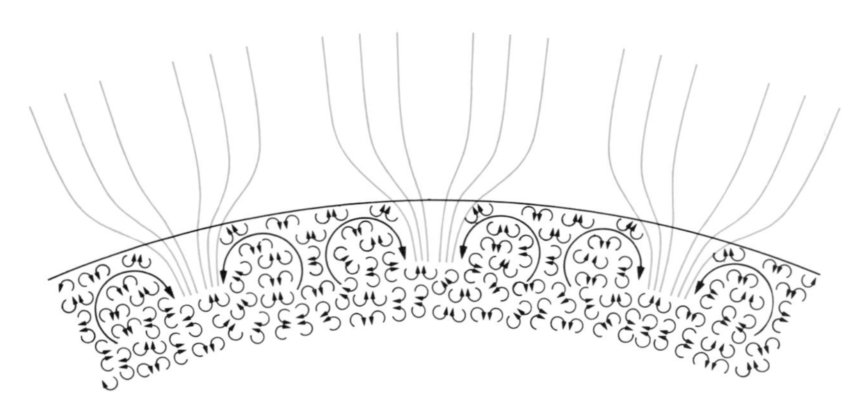

Figure 7.11. The horizontal motion of the supergranulation displaces the frozen magnetic field lines that both leave and enter the solar surface. They thus cluster together particularly densely where the material is sinking, i.e., around the edges of the cells. The motions are indicated by curved arrows; the small ones represent the granulation and the larger ones the supergranulation. In the centre of the supergranulation cells, where material is rising, no magnetic field lines are present. This explains why the solar-granulation pattern coincides with that of the magnetic fields

HERR MEYER AND THE YOUNG BOY AND GIRL

Subsequently, Herr Meyer often thought about the young girl and the freckled boy that he had unwittingly brought together. It was in the middle of a circus act in which a clown was trying to hang washing on horizontal magnetic field lines. He was holding a large sheet in a magnetic field. The plasma was then turned on. When he let go of the washing, it dropped slightly but remained suspended, hanging freely on the slightly sagging field lines. He proudly bowed to the applause of the crowd. But then when he tried to walk away he pulled the field lines with him, because he had also become plasma, and the sheet seemed to follow him. He went back to where he was before, and the sheet returned to where it had been hung. For several minutes the clown tried every way he could think of to escape from the sheet, but without success. If he tried to creep away slowly, the sheet followed him just as slowly. If, to fool the sheet, he looked thoughtfully at a woman in the first row, and then suddenly, without warning, ran away, it still followed him. Finally, he became tangled in the sheet.

However, Herr Meyer did not confine his attention to what was happening in the ring. He suddenly realised that the magnetic field lines reached him. They passed through a young girl sitting in front of him to his left before reaching his body. She had already attracted his attention, and was sitting next to a young man with a freckled face. What happened next is sketched in Figure 7.12.

Figure 7.12. While sitting in the circus tent, Herr Meyer twice scratched his head in puzzlement. As he did so, he forced the magnetic field lines passing though his hand to move. Waves travelled along the field lines and reached the girl's neck. The forces set up by the Alfvén waves produced by Herr Meyer tickled the back of her neck

Mr Tompkins had already noticed the field lines that ran through his body. 'Now we can study the motion of waves along the magnetic field lines in a plasma' he said, waving his hand backwards and forwards in front of him. When his hand was back in its original position, Herr Meyer could see how two crests, one on each side, were moving along the field line away from Mr Tompkins' hand.

'I don't understand that', he said, scratching his head. He had forgotten that one of the field lines passed through his left hand and also ran straight through the girl's neck. As he scratched his head, a wave travelled along the field line straight towards the girl in front of him. Herr Meyer knew that the field lines exerted a force. The girl must have been able to feel this. Because she was not wearing any glasses, she could not see the field lines, but she must have felt her neck being gently tickled. She obviously did not connect it with the magnetic field in any way, and thought it must be the young man making a timid advance. She gave him a friendly look, and he smiled back. They started to talk, and later Herr Meyer saw them walking across the field hand in hand. He would have liked to know the outcome.

ALFVÉN WAVES

In 1970, the Swedish physicist, Hannes Alfvén received the Nobel Prize for Physics. The prize honoured someone who had made a major contribution to our understanding of the properties of plasma. His greatest discovery was to find that a plasma containing a trapped magnetic field may initiate a form of wave motion that is not known in normal liquids or gases.

To understand this, let us imagine a plasma that contains what is known as a homogeneous magnetic field, one in which the field lines are parallel to one another, and where the field strength is exactly the same throughout the space (Figure 7.13a). Imagine that one region of the plasma, say that at the left-hand edge, is moved upward (Figure 7.13b). The field lines follow it, and thus become bent. The magnetic forces thus created tend, on the one hand, to pull the volume of plasma that has been moved back into its original position. On the other hand, the field lines in the regions on both side of the volume that has been moved attempt to shorten. As a result a kink in the field links propagates towards the right (Figure 7.13c, d). If the volume of plasma at the left-hand edge of the diagram is not simply moved, but returned to its original position, a wave-crest propagates towards the right (Figure 7.13e). If the plasma is moved backwards and forwards rhythmically, a whole train of waves moves towards the right (Figure 7.13f). These are the *Alfvén waves*. We know that waves exist in air, for example, and these are sound waves. However, they consist of changes in the air density. At each oscillation the air moves backwards and forwards in the direction of the waves' travel.

In Alfvén waves the material moves at right-angles to the line of

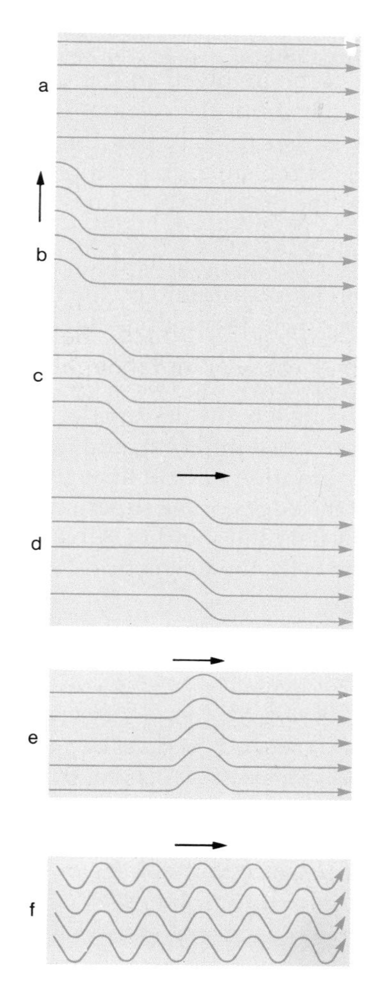

Figure 7.13. The propagation of waves along magnetic field lines. (a) Assume that the field lines in a plasma are straight and horizontal. If the plasma at the left-hand side is displaced upwards, the (frozen) magnetic field lines move with it. (b), (c), (d) The buckle in the lines propagates towards the right. (e) If the plasma is moved up and then down, a wavecrest propagates towards the right. (f) If the plasma is displaced up and down in a specific rhythm, a train of waves propagates towards the right: these are Alfvén waves

propagation. In Figure 7.13, the displacement of the field lines moves towards the right, while the plasma is moved up and down. Magnetic fields are responsible for this new form of wave. They do not exist in a plasma that is free from a magnetic field.

Alfvén waves appear to be involved in the calcium network. We have already mentioned that light from the calcium line is a sensitive indicator of temperature. We saw earlier (p. 120) that the magnetic network, which coincides with the edges of the supergranulation cells, is produced by the horizontal flow that drags the magnetic field lines to the edge of each cell. But why is the solar material in the higher layers hotter where the light from the calcium line is produced? The magnetic field lines concentrated at the edges of the cells are shaken and moved around by the rising and falling masses of gas in the granulation. This includes motions at right-angles to the field lines. These propagate upwards and downwards along the field lines in the form of Alfvén waves. Let us consider just the waves that propagate upwards. They travel into the outer layers, where the density decreases outwards. There the energy in the waves is converted into heat, and increases the temperature in the regions where a concentration of field lines rise from the surface, i.e., in just those regions where the edges of the supergranulation are found.

The concepts of frozen field lines and of Alfvén waves therefore explain why the supergranulation cells, the magnetic cells, and the calcium network all coincide.

8

TRACKING DOWN THE SOLAR DYNAMO

> For at least 2000 years astronomers have been attracted by solar activity: At first mainly the appearance and the variations of sunspots, later sunspot structure, prominences, coronal variations, eruptions with all their terrestrial consequences from aurorae to blackouts in radio transmission. . . . It is now known that all solar activity is a consequence of a magnetic field on the Sun.
>
> Michael Stix in *The Sun*

Since Hale discovered the Sun's magnetic field, we have known that nearly all phenomena observed—whether spots or flares, or even the various structures that (like prominences) appear only in the light of certain spectral lines—involve magnetic fields.

We still do not know precisely how the Sun's magnetic field originates, just as we are ignorant of the true cause of the Earth's magnetic field. Because both the gas in the Sun and the fluid interior of the Earth are plasmas, we have a few hints. We have already discussed the decay time of magnetic fields on the Sun, and the fact that the magnetic fields in sunspots do not originate or fade away with them. This led to the idea of magnetic 'ropes' hidden beneath the solar surface.

We shall soon see not only how these magnetic ropes originate, but also why they occasionally rise to the surface and produce a pair of spots. But first we need to discuss some other properties of a plasma.

THE ENERGY CONTAINED IN A MAGNETIC FIELD

Look at Figure 7.5 (p. 111) again. If we shift a layer of plasma towards the right—as shown there—the frozen magnetic field lines bend as indicated in the right-hand diagram. They are therefore forced to stretch. We have seen how they then exert a force on the plasma that tends to oppose the motion of the plasma. It takes considerable force to stretch the 'rubber bands' that are the field lines. If we move the plasma, we therefore need to expend energy. This energy is stored in the magnetic field. If we release the plasma, the field pulls the gas back to its original position. So magnetic fields are able to store energy. The more the 'rubber bands' are stretched and the closer they are forced together, the more energy is stored in the magnetic field, and the greater the forces that the magnetic field exerts on the plasma, tending to restore it to its 'preferred' state, where the field lines are as short and as far apart as possible.

Inside the Sun, energy is constantly pumped into the magnetic field. The cause is its differential rotation.

ROTATION AND THE SUN'S MAGNETIC FIELD

Let us assume that the Sun has a magnetic field, whose field lines leave the interior around the North Pole, curve far out into space, and re-enter the Sun around the South Pole.* The lines reconnect within the solar interior. In doing so, let us also assume that they do not enter the deep interior, but instead remain just beneath the surface. Initially, both inside and outside the Sun, they lie in a North–South direction, like meridians (or lines of longitude) on a sphere. The field is said to be meridional (Figure 8.1a). What happens as the Sun rotates?

If it rotated as a solid body, then the frozen magnetic field lines would rotate with it. But it does not. It takes points near the poles much longer to complete one rotation than it does a point in the equatorial region (see Figure 2.8). What does this mean for the lines of our magnetic field?

Because they are frozen into the solar material, they become wrapped around the Sun, as shown in Figures 8.1b and 8.1c. In doing so, the rotation must perform work against the magnetic forces that exist, because as the field lines are twisted up, they are stretched. The magnetic field gains energy and thus becomes stronger. We can see here how magnetic ropes are created like those we described earlier in discussing sunspots. After a large number of rotations, the field buried beneath the surface is tightly wrapped around the Sun. In the northern hemisphere it is directed towards the east, while in the southern it points towards the West.

Imagine that, for some reason, part of the magnetic field buried in a lower layer rises to the surface (Figure 8.1d). What will we observe? We would see two areas where strong magnetic fields break through the surface: in other words, a sunspot pair. In Figure 8.1d we can see that in the northern

---

* Here we abide by the convention that a magnetic-field region where lines leave the surface of the Sun is called a north magnetic pole, and one where they enter it, a south magnetic pole. In the example used here, the Sun's North Pole has north magnetic polarity. After 11 years the Sun's North Pole assumes southern magnetic polarity.

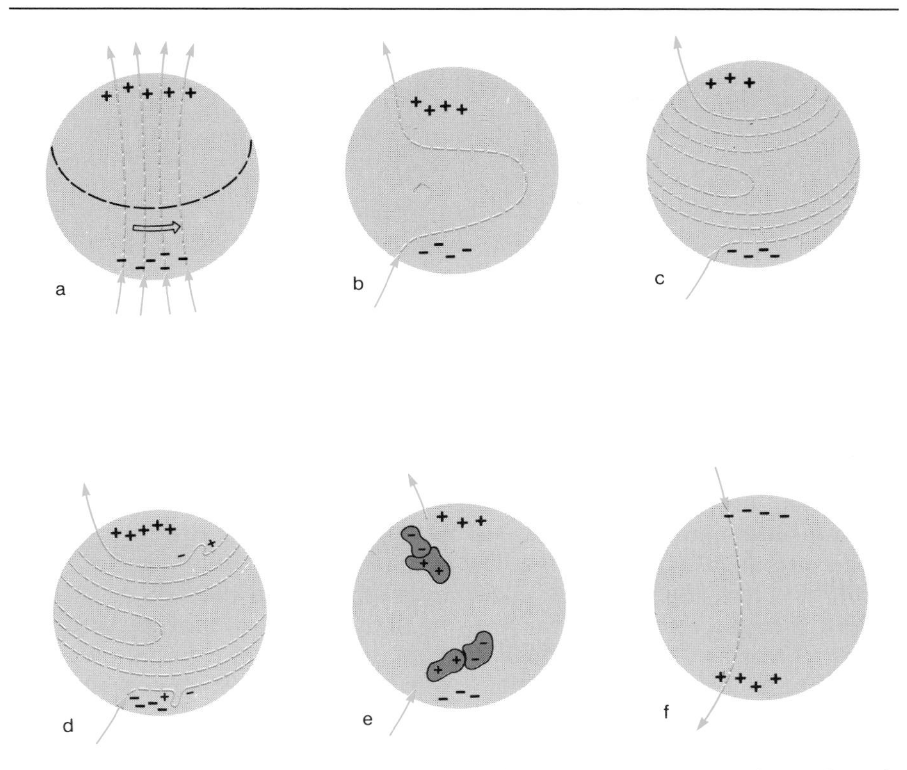

Figure 8.1. How the Sun's magnetic cycle, as shown in Figure 6.2, may be explained. (a) The magnetic field lines, which enter the body of the Sun and leave again in the northern polar regions, initially follow straight paths from South to North below the solar surface (dashed lines). (b), (c) Because of the faster rotation of the equatorial regions, those parts of the magnetic field that lie below the surface are stretched and pulled around the Sun. (d) The magnetic fields thus become stronger and stronger, and the magnetic forces cause portions of the magnetic 'ropes' thus created to emerge from the surface. As shown in Figure 7.7, bottom, this creates two magnetic spots. The field emerges from one, and re-enters at the other. (e) In the northern hemisphere the preceding spot has northern (+) polarity, and the following spot southern (−) polarity. In the southern hemisphere the polarities are reversed. (f) Through a mechanism that is still not understood, the signs of the polar field reverse, and we are left, after all the twisted ropes have come to the surface and faded away, with a field that corresponds to that shown in (a) but with reversed magnetic polarity. The six diagrams cover a single 111-year cycle

hemisphere the field lines emerge from the surface in the preceding spot, which therefore has northern polarity. The following spot, on the other hand, has southern magnetic polarity, because there the field lines re-enter the Sun. In the southern hemisphere the opposite occurs. The preceding spot has southern polarity, and the following northern polarity. As we have seen in Chapter 6, this is just what Hale observed.

We are beginning to unveil the secrets of the Sun's magnetic cycle. But we are still far from completely solving the puzzle. We have not yet established why the magnetic ropes hidden below the surface tend to rise through it. Above all, we do not know why the Sun's magnetic field regularly reverses. That does not seem to agree with our idea of magnetic field lines frozen into a plasma.

MAGNETIC BUOYANCY

Why should magnetic field lines lying below the Sun's surface rise and form sunspot pairs? The reason for this is that the magnetic ropes are buoyant. Let us consider a simplified case, where a magnetic rope is lying beneath the solar surface. This means that there is a tubular region of space that contains magnetic field lines, but that outside it there is no magnetic field (Figure 8.2). Because the field lines repel one another, they tend to cause the 'rope' to expand: the magnetic tube tends to increase its diameter because the magnetic field exerts an outward pressure. The plasma within the rope expands as the result of both its own internal pressure and that of the magnetic field. If no magnetic field were present, the pressure of the plasma in the rope would be equal to that of its surroundings, and the internal and external pressures would be in balance. The magnetic forces increase the interior pressure, however, and this produces an even greater tendency for the magnetic rope to expand. In doing so, the density of the material becomes less than that of its surroundings. The rope's buoyancy increases and it rises to the surface.

We have taken a single magnetic rope in an otherwise non-magnetic medium. In our picture of the way in which the magnetic field lines are twisted

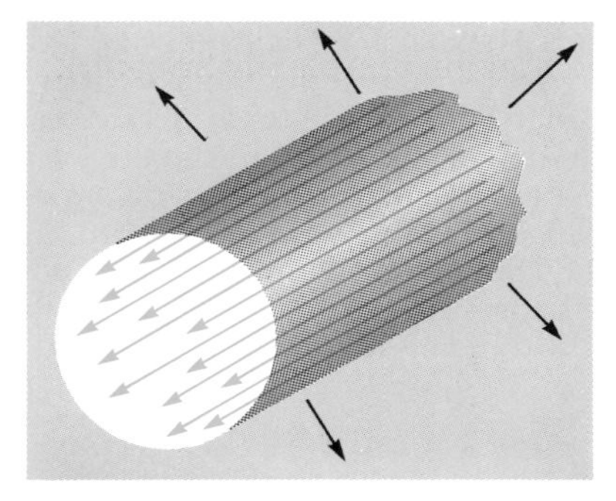

Figure 8.2. A magnetic 'rope' lying beneath the solar surface. Because the field lines attempt to mutually repel one another, there is a magnetic pressure outwards, which attempts to expand the rope. If there is to be equilibrium between the interior and the rope's surroundings, the internal gas pressure must be less than that outside. If the inner and outer temperatures are the same, then the density inside the rope must be less than that of the surrounding medium. The rope is buoyant

up beneath the Sun's surface, we have an entire magnetic layer, and not isolated ropes. However the granulation and supergranulation continuously agitate the twisted magnetic field lines. It suffices for a portion of the twisted magnetic field to receive a slight upwards thrust for it to enter a region where the magnetic field is weaker. Buoyancy then begins to act and the rope rises even farther. Although the magnetic field lines are slightly elongated by this process, this is of no particular importance, because the buoyancy effect is stronger than the force exerted by the lines and tending to pull it back down.

It almost seems as if we have solved the problem of sunspots. They occur when magnetic field ropes rise to the surface as a result of their own buoyancy. The fact that the first spots of a cycle appear at high latitudes, and that the spots get closer and closer to the equator over the following 11 years, must mean that as the field is gradually wound tighter, the magnetic buoyancy first becomes significant at high latitudes. The magnetic ropes at lower latitudes tend to rise later in the cycle. Ephemeral regions, however, which appear simultaneously with the spots, but at even higher latitudes and with opposite polarity, spoil this neat picture.

Even if we ignore the complication introduced by the ephemeral regions, we have still not understood why all the magnetic phenomena reverse their polarity in an 11-year cycle. It seems reasonable to say that an originally weak meridional field becomes wrapped around the Sun, and thus stronger, and produces pairs of spots through the effect of magnetic buoyancy. But what happens then? The events that we see happening on the Sun suggest that the next cycle begins just like the previous one, i.e., with a meridional field, but with opposite polarity. Towards the end of a cycle the magnetic field is tightly wrapped around the Sun, as shown in Figure 8.1. How can this give rise to a straight meridional field to begin the new cycle?

It should be stressed that we have described the solar plasma as being far simpler than it actually is. The conductivity of the solar plasma is not infinite. The field lines are not, therefore, completely frozen into place. We shall return to this point shortly.

HOW MAGNETIC FIELDS DECAY

Picture a plasma with a magnetic field, whose field lines are circles, like those in Figure 7.3a. We have already seen the way in which an electrical current creates such a field. As shown there, it passes through the centre of the circles, at right-angles to the plane of the page. Electrons are unable to move perfectly freely within a wire, however. They are not like a handful of grains of sand thrown out into weightless, empty space, which would continue to move in their original direction for ever. The electrons in a plasma repeatedly collide with ions, which deflect them, or even sometimes turn them back on their tracks. These constant interruptions to the electrons' motion through the plasma slow them down. The electrical current that they represent becomes weaker and the magnetic field produced by the current decays. This is why the magnetic field in a bar of copper decays when the latter is left to its own devices, and the field is not maintained by the constant application of currents from outside. We saw earlier that the decay time is greater, the larger the bar of copper, and that for double the thickness, the decay time was four times as long.

What happens when the magnetic field disappears? Do the magnetic forces simply become generally weaker, or do the field lines migrate out of the body? In this connection let us consider a field that, instead of having circular field lines, is like that shown in Figure 8.3. Here parallel, straight field lines running in opposite directions lie next to one another. How does this field decay? When the decay is examined in detail, it is found that the field first changes where neighbouring field lines run in opposite directions. It is as if neighbouring field lines cancel one another, just as opposite electrical charges neutralise one another. The magnetic fields are described as *annihilating* one another. The closer opposing magnetic fields lie to one another, the faster their mutual annihilation.

The decay of a field with circular field lines, like those traced by iron filings in Figure 7.3a, may also be considered as the annihilation of opposing magnetic fields. At diametrically opposite points on the circles the magnetic fields are opposed. The field therefore disappears first at the centre, because there the distance between opposing field lines is the smallest, and annihilation proceeds fastest. Fields with lines of a slightly greater radius subsequently disappear in sequence. The course of the decay with time is shown in Figure 8.4.

HOW TO GET ROUND THE DECAY TIME

If we apply the rule governing decay times to the Sun—as we did earlier for our bar of copper—we find decay times that far exceed the age of the Sun. If the Sun has a magnetic field, it must have possessed one since the very beginning. In addition, it certainly cannot decay. Our observations show that the Sun's magnetic field reverses every 11 years. Reversal implies that the field decays and a new one becomes established. According to our rule governing decay times, this ought to take millions of years. Where have we gone wrong?

We can circumvent the equation that describes decay times. Nature does, in fact, frequently do just that. We can see how to get round this problem by a simple thought experiment.

Imagine that we have two squares of equal area, each of which is traversed by parallel, straight, magnetic field lines. The field lines in one square are directed upwards and in the other downwards (Figure 8.5, left). How long will it take for the two opposing fields to annihilate one another? The first

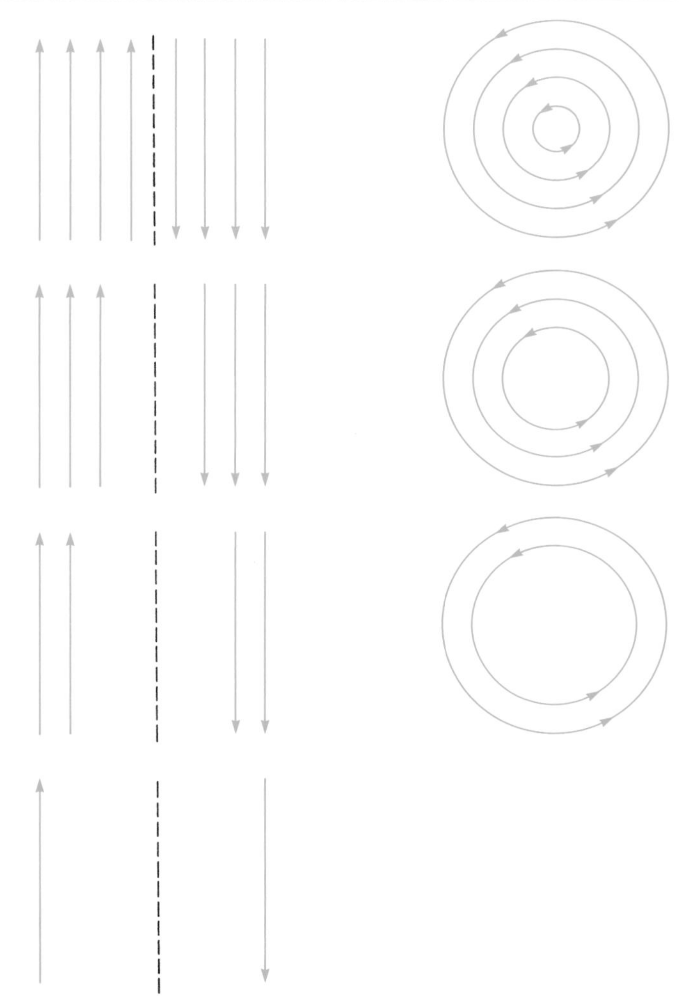

Figure 8.3. How opposing magnetic fields in a plasma annihilate one another, when the electrical conductivity is not infinite

Figure 8.4. The mutual annihilation of circular field lines in a plasma whose conductivity is not infinite

fields to cancel out will be those that lie next to the boundary between the squares. A field-free zone arises at the boundary and this propagates out into the two squares. If the length of the sides of the squares is one kilometre, and the plasma has the conductivity of copper, it will take a few years for the two fields to cancel each other out, because opposing field lines are, on average, one kilometre apart.

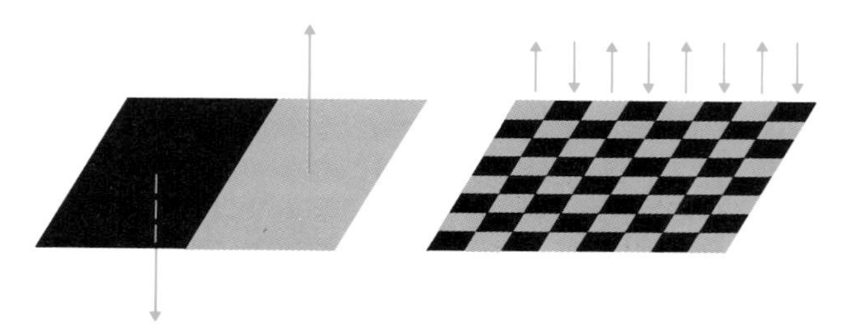

Figure 8.5. Two surfaces traversed by magnetic fields (left) are rearranged by horizontal motions to give a chess-board effect (right). In the lighter areas the fields are directed upwards and in the darker, downwards. Because opposing magnetic fields are closer in the right-hand diagram than they are in the left, they annihilate one another more rapidly

We can considerably shorten the decay times if we split up the columns of plasma, and their magnetic fields, keeping the magnetic field vertical, so that the two original squares become a form of chess-board (Figure 8.5, right). No force is required, because the field lines are not stretched, neither are they closer together. After this rearrangement, the decay times are shorter. If we assume that we now have 100 chess squares in each of our original areas, then the distance between field lines of opposite polarity is now one tenth of what it was previously. The decay time is one hundredth of its previous value, and now amounts to just a few weeks.

Anyone who wants to cause the magnetic field in a plasma to decay, merely has to bring opposing field lines close together. They then annihilate one another in a shorter time than the decay time corresponding to the original field.

HOW FIELD LINES UNTANGLE THEMSELVES

In nature, we find that magnetic fields frequently annihilate one another. Let us imagine magnetic field lines in a plasma as being like infinitely long spaghetti on a plate. If we begin to rotate a fork in our spaghetti-plasma, we wind the spaghetti into more or less solid clumps, which grow with every turn as we take up more and more of the pasta. In doing so, field lines of opposite polarity are brought close together, and thus annihilate one another. When this happens the twisted magnetic field becomes less complex. Figure 8.6 shows how field lines begin to react to twisting. If they are twisted even further, they will begin to annihilate one another. This is brought out very clearly in a computer simulation that has been produced by the Cambridge solar physicist, Nigel Weiss. Figure 8.7 shows a few individual frames from this simulation.

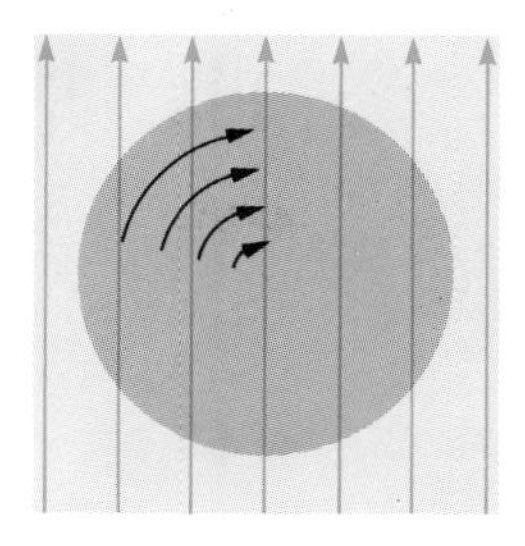

Figure 8.6. If a region of plasma crossed by a magnetic field is rotated (left), the (frozen) field lines follow the rotation and become 'twisted up' (right)

A circus artist who appeared in one of Herr Meyer's dreams made good use of this annihilation of opposing magnetic fields.

HERR MEYER AND THE ESCAPOLOGIST

'For the next act I need a volunteer to come into the ring,' cried the ringmaster, looking at Mr Tompkins and Herr Meyer, who were sitting in the first row. This was during the third show that they had attended. Herr Meyer plucked up his courage, and climbed over the edge of the ring. All eyes were upon him.

'Before I introduce Pedro, the world-renowned escapologist to this distinguished audience,' continued the ringmaster, 'I want to ask this member of the public to examine the apparatus.' Once again, the equipment consisted of two vertical metal plates, between which Herr Meyer, who was wearing his glasses, could see horizontal magnetic field lines. The ringmaster asked Herr Meyer to step inside the magnetic field and, because the plasma had not yet been switched on, he was able to do so easily. The field lines passed through his body, but he felt no effects. Then the lighting changed to red, and Herr Meyer knew that the plasma had been switched on. He could feel the force of the field, and he was no longer able to move his hands. If he tried to lift his arm by even the slightest amount, the lines followed it, and there was a strong force attempting to pull his arm back into its original position.

'Please step towards me', cried the ringmaster, and Herr Meyer would have willingly done so, because he did not feel very happy between the two plates. But however he tried, he was completely unable to move his legs. The force resisting him was so great that he could take no more than half a step forward, before he was pulled back. Herr Meyer found himself standing, completely helpless, between the two coils.

'Try and free yourself by using the rope ladder.' That was easier said than done. Herr Meyer, who had grasped the lowermost rung of the ladder that was hanging in the air above him, was only able to raise himself by a few

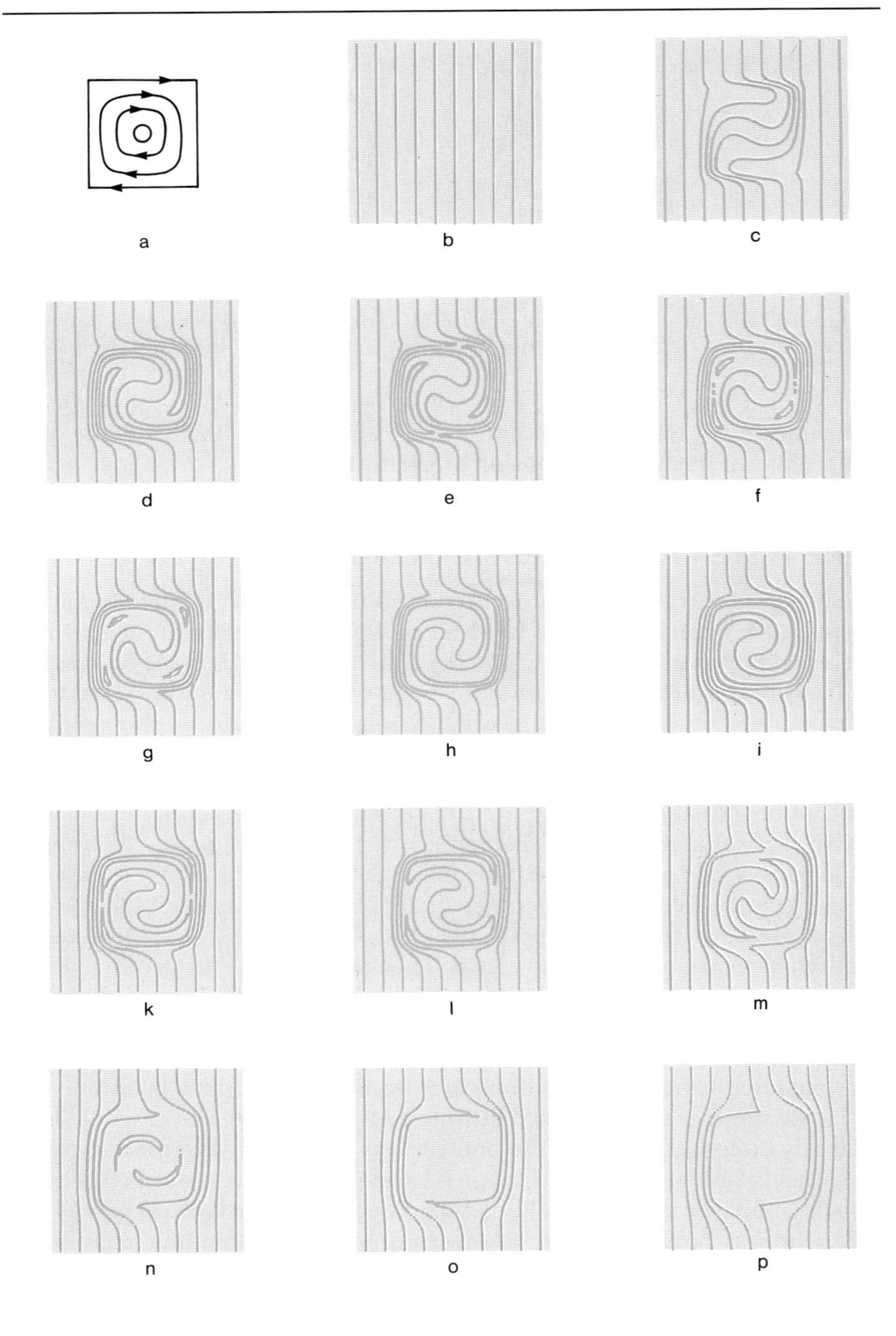
a
b
c
d
e
f
g
h
i
k
l
m
n
o
p

centimetres. The magnetic field was holding him down (Figure 8.8). Only when the ringmaster gave a sign and the plasma was switched off could Herr Meyer step free.

'And now, ladies and gentlemen, Pedro, the escapologist.' With these words, the man, dressed in a pair of shorts, who had been standing stock still and silent alongside the magnetic coils, and who had remained expressionless while the crowd were laughing at Herr Meyer's useless attempts to free himself, came forward. He bowed to the crowd and stepped into the magnetic field. The plasma was switched on again. For a moment the artist remained motionless, with his arms stretched above his head. Herr Meyer could easily imagine how he must be feeling the forces exerted by the magnetic field. Then the artist began to turn round. It was easy to see the amount of energy his legs had to exert to do so. He continued to turn until he had managed a half-turn.

Initially he had been facing Herr Meyer, so now his back was turned. He had dragged the magnetic field lines after him and they seemed to be wrapped around his body. Before long, he seemed to be finding it easier to turn. Eventually he completed a turn and began a second. Herr Meyer was not able to see exactly how it happened, but suddenly the field lines no longer seemed to pass through the man's body, but to encircle him instead. Now the artist was turning more quickly, first facing towards and then away from Herr Meyer in rapid succession, as if the magnetic field lines no longer had any effect on him. He grasped the ladder and effortlessly pulled himself up and out of the field between the two coils. The field lines were straight once more. A few field lines coiled round his body, but they were becoming visibly weaker.

The crowd burst into applause. The artist swung the ladder aside, jumped down beside the coils and took his bow.

The experience had a great effect on our two friends, and Herr Meyer at least, paid practically no attention to the rest of the performance.

'What happened', said Mr Tompkins, 'was that by turning round he managed to bring opposing field lines close together so that they annihilated one another. The field lines from the coils closed to form new lines that did not pass through his body. The lines that originally passed through him, closed around his body, and were no longer connected with the external field.'

Figure 8.7. How the field lines frozen into a plasma become twisted, break free of one another because of the finite conductivity, and re-form. The plasma rotates as shown by the black lines in (a), top left. A uniform magnetic field with parallel field lines (b) is twisted by the rotation, as shown in Figure 8.6. This causes opposing field lines to come together and annihilate one another. As a result, closed field lines that are no longer connected to the external field are created in the centre (m, n). Finally, the fields in the region that has been rotated disappear (o, p)

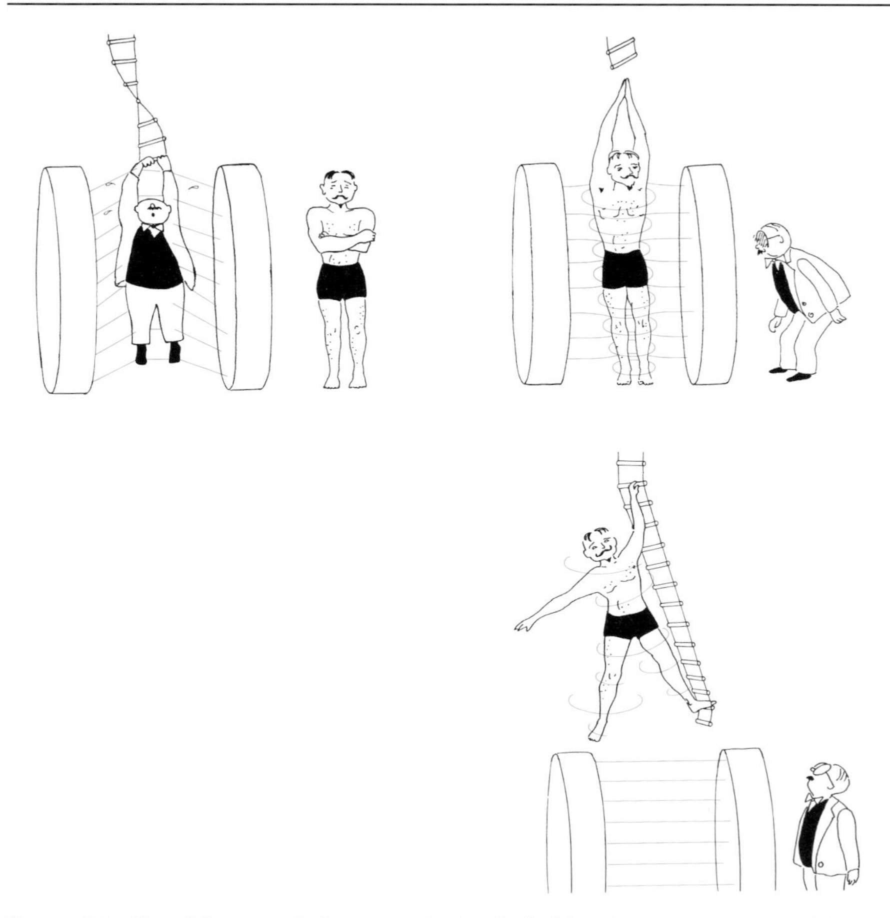

Figure 8.8. Herr Meyer and the escapologist. Left: Herr Meyer is prevented by the strength of the magnetic field lines from pulling himself up onto the rope ladder. Top right: The escapologist first rotated around his own axis. In doing so, opposing field lines are brought together, as shown in Figure 8.7, and annihilate one another. Right: The field lines that passed through the escapologist's body have broken free of those produced by the coils, and the artist can escape from the field with ease

Mr Tompkins' explanation was shown to be the correct one when, as they were going out, the artist came up to them and thanked Herr Meyer for his help. Mr Tompkins immediately asked him if the trick was not something to do with annihilation of the fields.'

'Yes, it is the mutual annihilation of field lines that earns me my living. It is no secret, because I perform in the open, not hidden behind a curtain or in a closed box. If it were not for the fact that field lines may mutually annihilate one another, I would not be able to escape.'

'But if it were not for that effect, I would never have had my accident 10 years ago, when I fell into the ring from a magnetic trapeze', he continued. As he spoke, Herr Meyer noticed for the first time that the artist dragged his right leg slightly.

'The act was later forbidden.' the artist said. 'Rather like the "Death-defying leap" that you saw earlier, I began the act by standing on a platform at the top of the tent, between two magnetic coils. The plasma was switched on, and I jumped downwards. Halfway towards the ground the field lines that I had carried with me braked my fall. After a few oscillations I hung suspended by the magnetic field about 10 metres above the ground. By rocking my body backwards and forwards I could start my magnetic trapeze swinging. It did not normally take me long to increase the swing until I was crossing from one side of the ring to the other. To make it less boring for the spectators, I used to swing above them with my feet and then with my head towards the ground. The act went down well, because people were impressed to see someone swinging through the air on invisible ropes. Especially because none of the special spectacles were issued for that act. The boss paid me a much higher fee than nowadays.

But then I had the accident. I may have been somewhat nervous that evening, or it may have been something else. I have often thought about it since, but have never been able to explain it. I had just reached my full swing. Then the magnetic field rope that was holding me up twisted. To be accurate, I should really say that I was hanging by two ropes. One came down from the magnet and entered my body on the left, while the other came out of my right side and ran up to the other coil. The two ropes therefore had opposing directions (Figure 8.9). That day, I somehow twisted my body sideways. The two sets of opposing magnetic field lines were brought close together, and they annihilated one another. The loop supporting me broke halfway up. I lost my link with the magnets at the top of the tent and fell into the ring. When I finally finished with doctors, I developed my escape act. The magnet is the one that I used previously, but now the magnetic field is stronger.'

THE REVERSAL OF THE SUN'S MAGNETIC FIELD

Let us return to the Sun's magnetic cycle. If we start with a field running North–South, as shown in Figure 8.1a, we can understand how buoyant loops in the twisted magnetic field lines create sunspot pairs in both hemispheres that have the observed polarities. We have also seen that there are reasons why the field lines tend to rise to the surface, when their magnetic fields are sufficiently strong. The basic cause of the solar cycle would no longer be a secret if we only knew what else occurred. We expect the field to become even more tightly wrapped around the Sun, and for any new sunspot pairs that appear to always have the same polarity as earlier pairs. Yet this is true

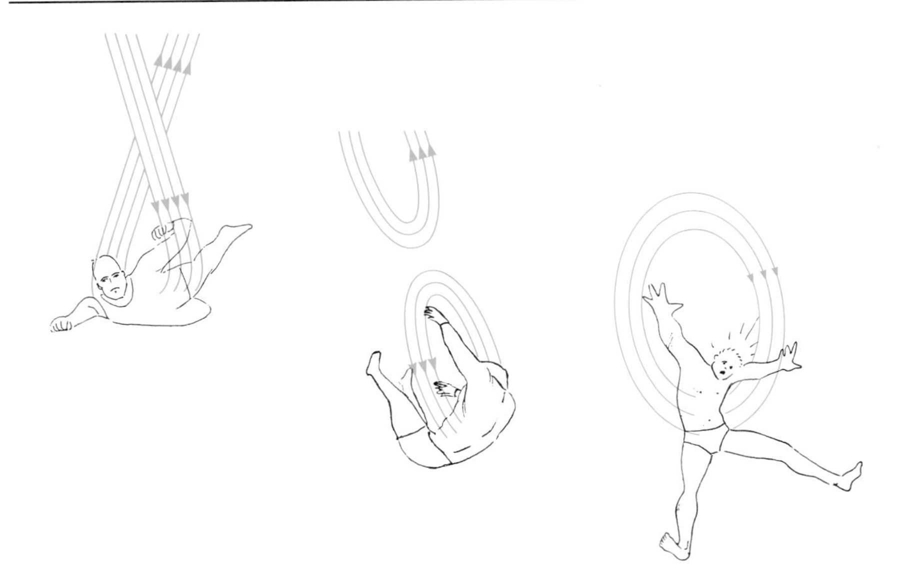

Figure 8.9. The circus accident. Left: The artist, who normally swung from the magnetic field lines from the magnet at the top of the tent as if from a trapeze, accidentally twisted his body. This brought the two halves of the loop of field lines that was supporting him dangerously close together. Because of the plasma's finite conductivity, the field lines were able to annihilate one another, and the magnetic field that ran through the artist's body became separated from the field originating at the top of the tent (centre). The field lines passing through his body were no longer supporting him, so he fell into the ring (right)

only if we adhere strictly to the idea of frozen field lines. We have, however, already seen that this begins to lose its validity if field lines start to mutually annihilate one another. We must expect this to occur in the outer layers of the Sun, however, where the granulation and supergranulation move the field lines around in a complicated fashion. This repeatedly brings oppositely directed field lines together, allowing them to annihilate one another. As yet we do not fully understand the interaction between turbulent solar material and magnetic fields. There are, however, signs of a suitable theory, which was been developed by Fritz Krause and Max Steenbeck (1904–81) in what was then East Germany (in particular), where the strongly twisted magnetic field is annihilated through turbulence in the solar material, leaving a weak North–South field, similar to that shown in Figure 8.1a, but with an overall field reversal. The process can then begin again. The solar rotation winds up the field lines once more, and everything follows the same pattern as 11 years previously. But now all the field directions are reversed, just as has been observed since Hale's original discovery.

We therefore have an approximate idea of how the solar cycle originates. There are, however, other magnetic phenomena on the Sun that we can explain more or less plausibly with our simplified concept of a plasma. One of these is that of filaments and prominences, another processes in the corona, and yet a third the bright flares that suddenly and unexpectedly erupt on the surface of the Sun. We shall describe these phenomena in the following sections.

PROMINENCES

Magnetic forces also seem to be responsible for enabling prominences to hang, like sheets of paper standing on edge, high above the Sun without collapsing down onto the surface. The key to understanding this was the observation that filaments nearly always occur above the boundary between extended areas of different polarity. Figure 8.10, top left, shows two areas of opposite polarity. The diagram (top right) shows the path of the magnetic field lines, as seen from the side. The lower diagrams show, schematically, how a filament occurs along the boundary between the two areas of opposite polarity. The filament appears to lie inside the coronal gas, just where we expect the field lines to be horizontal. This suggests that the denser material in the filament, which is subject to the Sun's gravity, bends the field lines downwards (Figure 8.10, bottom right). The compressed field lines exert an upwards force, which balances the gravitational pull. The filament hangs suspended in the magnetic field. In a world of plasma, magnetic field lines may serve as washing lines.

In the 1950s, Arnulf Schlüter, who, until recently, worked at the Max Planck Institute for Plasma Physics at Garching near Munich, and I were able to link the way in which filaments seem to hover above the Sun's surface to invisible magnetic fields. Since then almost four decades have passed. It was later realised that the image of filaments hanging suspended in the magnetic field lines is undoubtedly too simplistic. Observations clearly show that within them, material streams downward, and this does not seem to correspond to our view of frozen field lines. We are therefore dealing with properties that may only be explained by a more complicated model.

We have already seen that occasionally a prominence is ejected away from the Sun. Magnetic forces cause this to happen, because a magnetic field is capable of storing energy. If a plasma is moved in such a way that the field lines frozen into it are stretched or twisted, work has to be expended. A magnetic reservoir of energy may be created by greatly extending a magnetic field in its longitudinal direction, and compressing it in the perpendicular direction. If it becomes possible for that energy to be suddenly released, the plasma may be accelerated to high velocities. We shall see later that the stored energy may also be converted into heat.

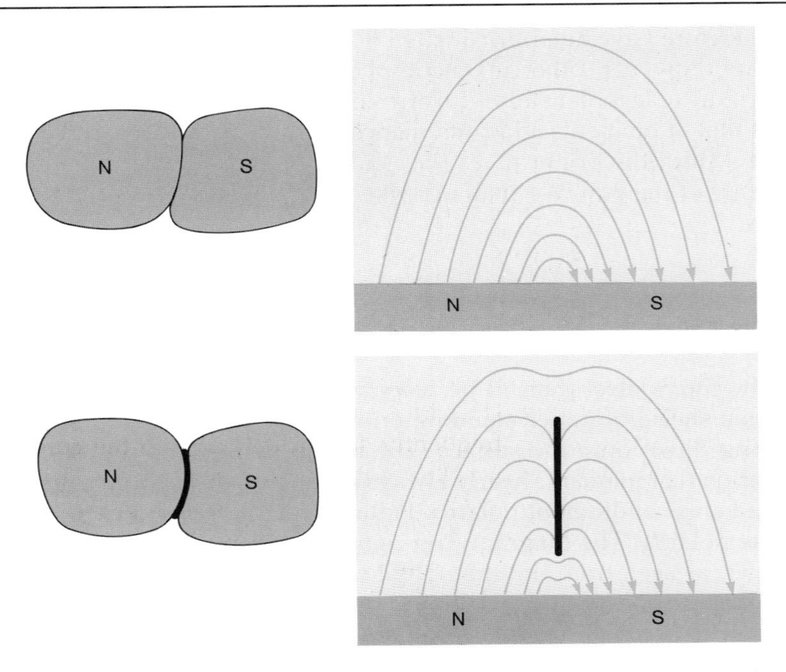

Figure 8.10. How prominences are kept suspended in the solar atmosphere by magnetic fields. Top left: Two regions of opposite polarity lie alongside one another. Top right: The magnetic field between the two areas, as seen in section. Below left: Hydrogen gas collects at the boundary between the two magnetic regions, with magnetic fields frozen into the gas. Below right:The cloud of gas is dragged downwards by gravity, carrying the field lines with it, causing them to curve downwards, and forcing the hydrogen to collect in a thin, vertical layer. The resulting forces counteract the gravitational drag, holding the gas suspended above the Sun's surface as a thin sheet

MAGNETIC FIELDS IN THE CORONA

Occasionally, glowing clumps of material from a prominence are seen to shoot sideways across the face of the Sun. Their paths enable us to determine the tracks of the otherwise invisible magnetic field lines in the corona. Because the fields are frozen into the visible clumps of gas and are extremely strong, the material is forced to follow them and is unable to react to the force of gravity by dropping straight down to the surface. Like chairs on a ski-lift, the clumps of material are forced to slide downwards at a shallow angle.

The corona is permeated by magnetic fields. In particular, there are loops, along which glowing material may be seen to flow, and which begin and end at the surface. Their shape (Figure 5.3) may be confidently interpreted as of magnetic field lines. We have learned more about magnetic fields in the corona from X-ray radiation.

Solar X-rays have been familiar to us ever since we have been able to study the Sun with instruments deployed above the Earth's atmosphere, which the X-rays are unable to penetrate. We shall discuss some of the results obtained from satellites and space stations in Chapter 12.

Most of the Sun's X-rays come from the corona, which, because of its temperature of millions of degrees, 'glows' at X-ray wavelengths. It appears like a thin, glowing layer of mist above the Sun's surface, which is relatively dark in X-rays. The corona's X-ray luminosity is not even, however. There are brighter patches, occasional bright knots, and even dark regions that hardly emit any X-ray radiation. The latter are known as *coronal holes,* and were first discovered by space probes (see Figure 12.2).

In fact, the density of the hot coronal gas within them is far less than in other regions of the corona. The cause of this lies in the magnetic field. The magnetic field lines, which frequently stretch far out into the corona, are rooted in the solar surface. Areas of northern and southern polarity often coexist peacefully alongside one another. The field lines leave one of the two magnetic regions, and return at the other. The shape of the field lines then resembles that shown by the iron filings slightly above the centre of Figure 7.8 left.

Frequently, however, regions of similar magnetic polarity lie alongside one another. Their field lines then resemble those around two identical magnetic poles, somewhat as seen above the centre of Figure 7.8 right. Figure 8.11 shows schematically how the two forms of field lines may occur within the corona. With opposite polarities, the field lines are confined fairly closely to the Sun, but when the polarities are the same the lines may extend far into space.

Imagine that the corona is evenly filled with a thin plasma. In the areas marked 'a' in Figure 8.11, where field lines both leave and return to the surface, the coronal plasma cannot escape. The field lines are forced to remain frozen into the plasma. If the material were to move away from the Sun, the field lines would have to lengthen. Their elasticity would prevent any attempt by the material to escape. The situation is very different in the regions marked 'b'. There the gas can escape into space along the field lines, without the lines being either extended or compressed together. The magnetic field does not offer any obstacle to movement of the material. Anywhere that the magnetic fields have a structure like that at 'b' in Figure 8.11, the corona tends to empty. It is not, therefore, surprising that there is no radiant material in such regions, and that we detect no X-rays from these coronal holes.

The truth of this view is based not just on the independently observable structure of the magnetic field, but also on other evidence. The Sun rotates, and carries the coronal magnetic fields that are frozen to the surface with it, taking about 27 days to complete one rotation. Seen from the Earth, the coronal holes also move across the disk of the Sun. When we see one of the holes close to the centre of the disk, it is 'aimed' straight at us. The outwardly

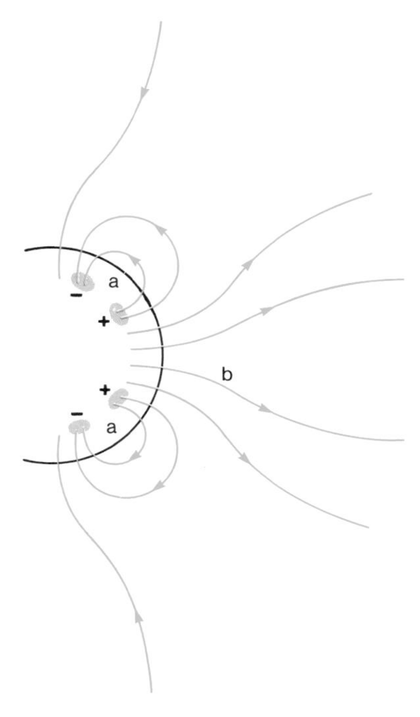

Figure 8.11. Magnetic fields in the corona. In the regions marked a, the field lines are closed, and trap the plasma. In the region marked b, the lines extend into space, and the plasma may escape along the lines. Such a region is known as a coronal hole

flowing coronal material is moving in our direction. The solar wind, which is responsible for a constant loss of coronal material to space, reaches us in about three days. From the difference between the time at which the hole was aligned in our direction and that at which the blast of wind hits us, we can determine the velocity of material that leaves the Sun through coronal holes as about 600 km/s.

What indicates that we have become the target for material from a coronal hole? About three days after one has crossed the centre of the Sun's disk, we observe disturbances in the Earth's magnetic field. It begins to 'oscillate'. The reason for this is the solar-wind plasma, which has encountered the Earth's magnetic field, and is affecting the terrestrial field lines.

DETACHED FIELD LINES

Imagine we remove a ball of plasma from a magnetic field, as we have done earlier in Figure 7.6, right. The field lines attempt to follow the region of plasma, because they are frozen into it. In doing so, however, field lines of opposite polarity are brought into close proximity (see Figure 8.12). The field lines of the ball of plasma separate from those that are anchored in the stationary plasma, giving rise to one set of field lines that accompanies the moving ball of plasma, and another that remains behind with the stationary portion of the plasma.

Although this was just a thought experiment, it does also apply in practice. Let us reconsider the situation where some coronal material is bound to the Sun by magnetic fields, and some leaves the region controlled by the Sun's influence. We saw that material held by closed field lines could not escape from the Sun, whereas material around open field lines may flow out into space. However, even the material trapped by the field lines anchored in the Sun may occasionally escape. Such pressures may build up that the field lines may be forced out to considerable distances from the surface. Lines of opposite polarity then lie alongside one another and may mutually annihilate one another. The lines reconnect to give a magnetic-field region that is no longer anchored to the Sun, to which the plasma is no longer bound. The ball of plasma has become detached from the Sun, and can escape into space.

FLARES

Energy is stored in a plasma's magnetic field. We only have to think of a twisted rope of magnetic field lines to see that a magnetic field on the Sun contains energy. In one cubic kilometre of plasma from the centre of a sunspot there is as much energy as was released in the atomic bomb that devastated Hiroshima. What becomes of the energy when two magnetic fields of opposite polarity annihilate one another? Figure 8.13 shows a sunspot group in white light (left), and as a magnetogram (right). We can see that the left-hand spot has northern polarity (white) and the right-hand one southern polarity (black). Particularly noticeable, however, is a smaller spot, below the right-hand one. The magnetogram shows that above it, fields of opposite polarity are tightly pressed together. The boundary between the white and black regions is very sharp. Shortly after these images were taken, a flare erupted here. Subsequently, hardly anything could be seen of the magnetic fields in the area, because they had mutually annihilated one another.

Flares seem to derive their energy from the annihilation of magnetic fields. Just as the electrical energy stored in a thundercloud produces a flash of lightning, the magnetic energy in a portion of the Sun's magnetic field may be released in a very short space of time—which as we have seen may be just a few minutes—as a flare. In both cases, electrons are accelerated to extremely high velocities. Lightning strikes in the Earth's atmosphere and solar flares are significantly different, however. In a flash of lightning the light comes from heated gas, because the air suddenly becomes conducting

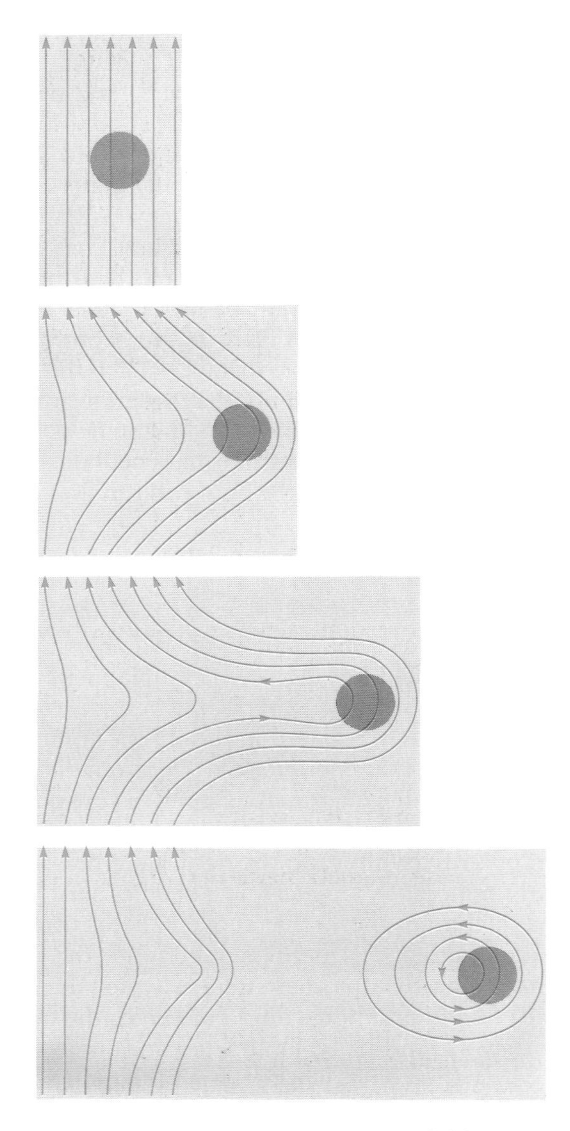

Figure 8.12. If the experiment shown in Figure 7.6 (right) is repeated with a plasma having finite conductivity, then the opposing field lines annihilate one another, and the ball of plasma becomes detached from the original magnetic field

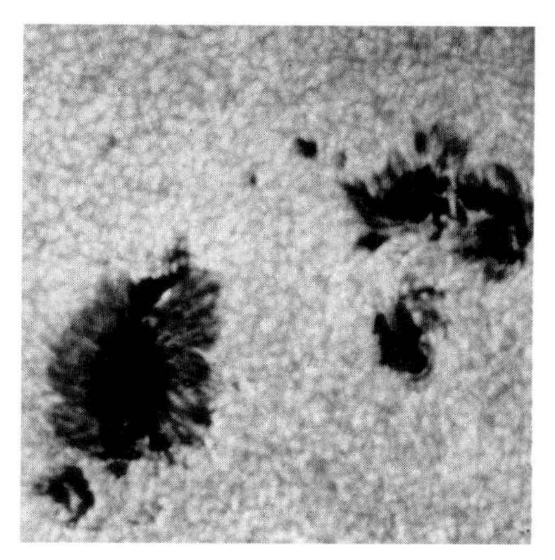

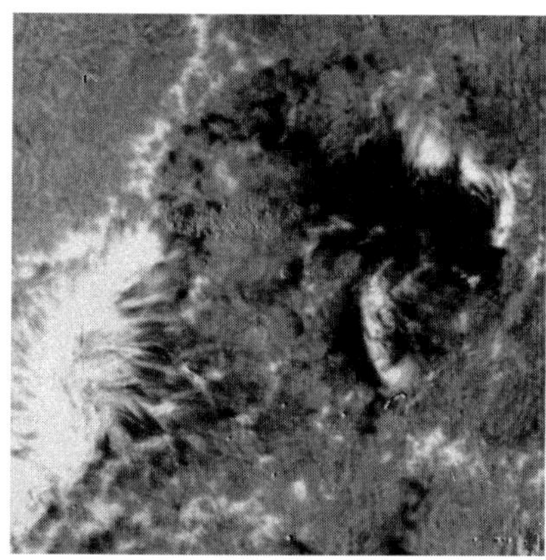

Figure 8.13. Left: A sunspot group in white light. Right: The magnetogram of the same area shows that the left-hand spot is of northern polarity (white), and the other at top right has southern polarity (black). The smaller spot below it exhibits a sharp boundary between opposite polarities. (There is a sharp border between black and white.) There the fields were able to annihilate one another very rapidly. Shortly after these images were obtained, the energy was released in a solar flare (photos. Lockheed Research Laboratory, using the Swedish Vacuum Telescope on La Palma)

and, as in a short circuit, the current, which can suddenly flow unhindered, produces extremely high temperatures. Electrical energy is converted into heat. In a flare, magnetic fields annihilate one another, and the electrons in the plasma reach speeds comparable with that of light. The extremely high-velocity electrons, produced by the gradual strengthening and then the sudden annihilation of magnetic fields in a flare, give off radiation, known as *synchrotron radiation*, which we detect as light.

We believe that the energy of a solar flare comes from the magnetic fields. From what, however, do the magnetic fields derive their energy? It is thought that it is the rotation of the solar surface, with a rapidly rotating equatorial zone and more slowly rotating polar regions, that twists the field lines and thus stores energy in the field. Again, however, where does the Sun's rotational energy come from? We do not know. We have already discussed the granulation, which keeps the outer layers of the Sun in constant motion. In the next chapter we shall see how it is produced by the flow of energy from the solar interior. These motions act in such a way that the Sun does not rotate as a rigid body. If this is all correct, in the final analysis it is the nuclear energy in the centre of the Sun that is responsible for the repeated, 22-year, magnetic cycle that we observe.

This is why we now need to consider the processes that are occurring in the deep interior of the Sun.

9

THE SUN INSIDE A COMPUTER

> Amid this great population the sun is a humble unit. It is a very ordinary star about midway in the scale of brilliancy. We know of stars which give at least 10,000 times the light of the sun; we know also of stars which give 1/10,000 of its light.
>
> Sir Arthur Eddington in *The Nature of the Physical World*, 1930

Despite the arsenal of instruments and methods of measurement at the disposal of modern solar researchers, in one respect we are no better off than Fabricius, Scheiner and Galileo: Even today we cannot actually see any farther into the body of the Sun itself. The best vacuum telescopes and the solar observatories in orbit around the Earth are just as restricted. Although solar material is not completely opaque, only light from the very outermost layers reaches us directly. Even with our most sophisticated optical telescopes, we are still looking at only a thin superficial layer of the Sun. The very largest telescopes offer no better view of its interior than did the telescopes that were used by the earliest solar observers.

The fact that, despite this, we know more about it than we do about the interior of the Earth is because even the material at the very core of the Sun has far simpler properties than the materials that form the Earth. This again is primarily a result of the high temperature of the solar material. Within it, atoms have long ago lost most, or even all their electrons. Ions and electrons intermingle freely. They are unable to form complicated molecules or crystals. This makes life much easier for those of us who are trying to fathom what happens within the body of the Sun, without being able to actually see down into the interior.

Spectral analysis indicates what materials are found in the Sun. The mean density may be calculated from the Sun's mass and its diameter. The Sun contains about 300 000 times the mass of the Earth. It is roughly 100 times the Earth's diameter. As a result, its mean density amounts to about 1.4 times that of water. In contrast, the density of the Earth is about 5.5 times that of water. At a relatively early stage the temperature at the centre of the Sun was estimated as a few million degrees, which is a value that is not very far from what is accepted nowadays. We are well-informed about the properties of matter at such relatively low densities and extremely high temperatures.

When computers became available after the Second World War, it finally became possible to solve the equations describing the structure and evolution of stars, and thus—at least to a certain extent—simulate stars inside a computer. This has helped us to understand stars and their life histories, and in particular it has revealed the history and internal structure of the Sun.

NUCLEAR FUSION INSIDE THE SUN

As we have already seen in Chapter 1, the Sun's energy arises from its atomic nuclei. Physicists such as George Gamow, Hans Bethe, and Carl Friedrich von Weizäcker made major advances in our understanding of what takes place. It is the so-called *proton–proton chain* that is primarily responsible for the Sun's energy.

The three stages of this process are schematically shown in Figure 1.2: As a result of the extremely high temperature, two hydrogen nuclei, i.e., two protons, collide at high velocity. Protons have a positive electrical charge. When they encounter one another, the repulsive force created by their identical charges tends to either brake or deflect them. Only very rarely do they come close enough to combine. Thanks to an effect of quantum mechanics that was discovered in the 1920s, however, at any given instant enough protons do combine inside the Sun to form new atoms of deuterium, which has been mentioned on p. 7. The positive electrical charge of one of the protons is emitted by the new atomic nucleus in the form of a positron. This soon encounters one of the abundant free electrons, and the two annihilate one another to create a quantum of radiation. If the new deuterium nucleus encounters an ordinary hydrogen nucleus, i.e., a proton, the two combine to form yet another new atomic nucleus. This process, which is shown (centre) in Figure 1.2, turns one nucleus each of ^{1}H and ^{2}H into a nucleus of ^{3}He (using the notation described in Chapter 1). This step also releases radiation, which leaves the site of the collision, and increases the energy of the free electrons flying around in the neighbourhood.

The ^{3}He isotope is not very common inside the Sun, because, hardly has it been formed when, in the process shown in the third diagram in Figure 1.2, two nuclei collide to create another form of helium, ^{4}He. As the diagram shows, two protons are also released. Overall, therefore, four protons are turned into a single helium nucleus. The mass of a helium nucleus, however, is slightly less than the total mass of the four protons from which it was created. About 0.7 per cent of the original mass is missing. Part was lost as a positron, which then almost immediately collided with an electron and turned into radiation. Another part was emitted as a quantum of radiation in the process that created the ^{3}He nucleus. Mass is turned into energy as hydrogen is turned into helium. This is the mechanism that converts mass into energy within the Sun. In fact, another energetic particle, a neutrino, is released when the deuterium nucleus is formed. We shall return to this point later.

We have, however, rather oversimplified the description of the processes involved. As well as the processes shown in Figure 1.2, there is a set of secondary processes, and these are shown schematically in Figure 9.1. They have little significance in the Sun's overall energy balance, but they do cause some headaches among astrophysicists. Again, we shall return to this point later.

THE BIRTH OF THE SUN

If we want to simulate the Sun's evolution in a computer, we first need to ask the question of how we should begin. How did the Sun begin? Has it always been the same as it is today? That cannot be true, because it must possess a finite supply of energy. It cannot have been radiating as it is now since time immemorial. So how was the Sun created? Other stars offer us some clues to the answer. Astronomers know that high-luminosity stars have short lives. The brightest stars in our Milky Way system are the youngest. Many formed only recently, even though 'recent' in this context should be taken to mean (say) a million years. For stars that is a very short period of time. Even though this span of time may seem inconceivably long, we can compare it with the period over which *Homo sapiens* evolved. A million years ago there were still ape-men in Java.

The areas of the sky where stars have recently formed provide us with information about how the process takes place. We always find numerous gas and dust clouds, remnants of the material that lies in the space between the stars (known as interstellar matter), from which they have obviously been born. Time after time, clouds of gas collapse under the influence of their own gravity. Concentrations of mass occur from which stars arise. These are always born in groups and such groups of stars formed from the collapse of a gas cloud are seen in the sky, and are called *clusters*. They often contain a few hundred, or even as many as 100 000, stars that in all probability formed simultaneously. The Sun must also have arisen in the same way, and somewhere in the Milky Way there are undoubtedly still hundreds of its siblings. In the intervening 4.5 thousand million years they have become separated from one another and spread over a large region of the Galaxy. We are unable to distinguish them, however, from the thousands of millions of other stars.

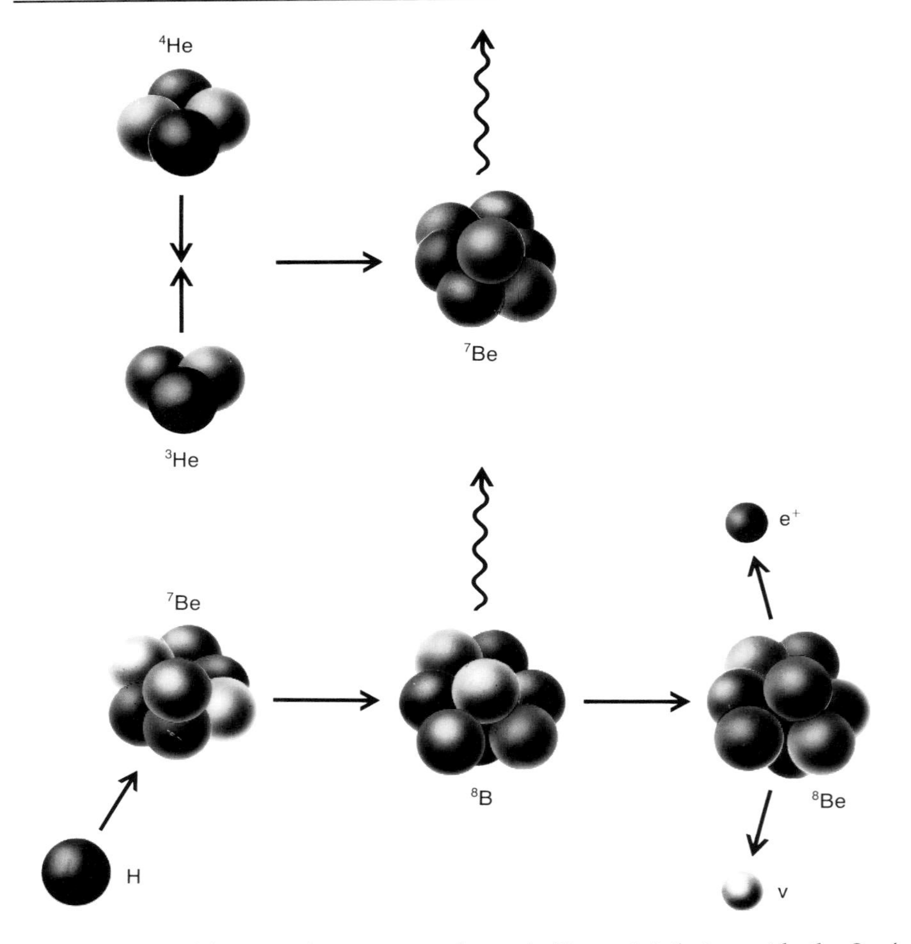

Figure 9.1. In addition to the processes shown in Figure 1.2 that provide the Sun's source of energy, there are several secondary processes that are of little importance in the energy budget. In the reactions shown here, boron atoms (^{8}B) are formed from helium and hydrogen atoms, and these then later decay into beryllium (^{8}Be). This process is important because it creates high-energy neutrinos (v). They escape unhindered from the centre of the Sun and reach the Earth, where they may be investigated

Our picture of the formation of the Sun is still incomplete. Luckily, the Sun's subsequent history does not depend on the specific details of the processes by which it was born. The same often applies to people, of course. The fate of a healthy baby does not depend on whether it was born naturally or as the result of a Caesarian operation.

In any case, following the collapse of the initial cloud, we are left with a sphere of gas held together by its own gravity. The gas became so hot during the last phase of its contraction that the pressure prevented it from collapsing to a single point. A condensation formed at the centre, where gravity, which attempted to cause the sphere of gas to collapse, and gas pressure were in balance. Although the material within the sphere had nearly come to rest, the remnants of the cloud continued to rain down onto its surface (Figure 9.2). It did not take long for its central temperature to reach about 10 million degrees. At that temperature, hydrogen ignites: the nuclear reactor at the centre of the sphere of gas had been switched on. The chemical elements in the gas were still those of the original interstellar material. Whatever its earlier history may have been, the final sphere of gas is unaffected by it. This is lucky for research, because when we want to study the later history of the Sun, we can start our computer models at this phase, without having to know anything about its previous history.

SEARCHING FOR THE PRIMORDIAL SUN

So we begin with a sphere of gas with the mass of our Sun. But what sort of materials must we include? The most common elements in the universe are hydrogen and helium. Nearly all stars consist of a mixture of these two elements. Only very small amounts of other elements are ever present. Quantitatively, they are insignificant and merely add a bit of spice to the hydrogen–helium mixture. Unfortunately, it is very difficult to detect helium in stellar spectra, so we cannot be absolutely sure of the exact hydrogen/helium ratio in the Sun. It appears that the most probable value is about 7:3 by mass. We shall shortly see how we can try to determine the original mixture of the two elements from current observations, i.e., made some thousands of millions of years later.

From determinations of the age of the Earth and meteorites using radioactive elements, we find that the age of the Solar System is 4.6 thousand million years. With this we can begin to experiment. We take as our model for the primordial Sun one with initial composition ratios of hydrogen to helium to other elements of 70:29:1. In our computer we follow how, in the inner core, hydrogen is turned into helium, releasing energy, which causes the Sun to shine. We find that the primordial Sun radiated about 40 per cent less energy than today, and it was also somewhat smaller. Over the course of time part of the hydrogen at the centre has been converted to helium. Simultaneously, the Sun has become hotter and larger. We find, however, that with the original ratios of elements that we chose, even after 4.6 thousand million years, our model would still not have the Sun's current luminosity. The reason for this is that we chose incorrect values for the initial mixing ratios. The error is easy to correct, however, and so we can alter the ratios of hydrogen, helium, and other elements that were originally present. After many trials, we obtain a model for the early Sun, which, 4.6 thousand million years after it began as a chemically homogeneous star, corresponds to the luminosity of our Sun today. Fairly recently, the astrophysicist Achim Weiss at the Max-Planck Institute for Physics and Astrophysics in Garching calculated solar models. He was able to reproduce all the properties of our current Sun when he began with mixing ratios of 700:286:14.

Using this method, it is not just possible to calculate the history of the Sun from being a protostar to its current state. A model of the actual Sun also allows us to take a look at its interior, a pleasure that is otherwise denied to solar researchers. Figure 9.3 shows a schematic picture of the interior of the Sun, as determined by our computers. In the past 4.6 thousand million years the amount of helium at the centre has grown: where there was once only 286 grammes of helium in every kilogramme, this has now increased to about 700 grammes. Our model of the Sun also shows how the intermediate and secondary products of the proton–proton chain are distributed within the solar interior.

A glance into the interior of our model Sun shows that the energy released by the nuclear processes is primarily transported to the exterior by radiation. When, during their continuously zigzag flight, electrons and atomic nuclei occasionally pass close to one another—being repeatedly deviated by the electrical fields that surround these charged particles—they emit innumerable, small electrical discharges, photons (or quanta of radiation), which we have discussed in Chapter 3. The photons are repeatedly deflected and scattered by the mixture of electrons and atomic nuclei inside the Sun. Any individual photon has an exceptionally difficult task in reaching the surface. The longest straight path that it may cover is no more than a few millimetres. It then encounters another electron or an ion and is deflected from its original track. It is repeatedly reflected back into the layer that it has just left. With this constant hither and thither, backwards and forwards, its progress towards the exterior is painfully slow. The time that it requires to complete its journey to the outside world is measured in millions of years. A

Figure 9.2. How we believe the Sun formed from a gas cloud. (a) A cloud of gas, like those found today between the stars, began to contract. Initially, the density is the same throughout the cloud. (b) After 390 000 years, the density in the centre of the cloud has increased 100-fold. (c) 423 000 years after the beginning of the process, a hot core forms in the centre, which—at least initially—ceases to collapse. It is shown magnified on the right. Its density is now 10 million times the initial value. The major part of the mass, however, remains in the surrounding, and still contracting, gas cloud. (d) Shortly afterwards, when hydrogen molecules disintegrate into individual atoms, the core recommences its collapse and forms a new core (shown enlarged at two different scales), which is already roughly the size of the Sun. Although at this stage it is not very massive, all the material in the cloud will collapse onto it over the course of time. The core becomes so hot at its centre that hydrogen begins to fuse and a star is born, which only differs from our Sun in that it is just beginning to tap its reserves of nuclear energy

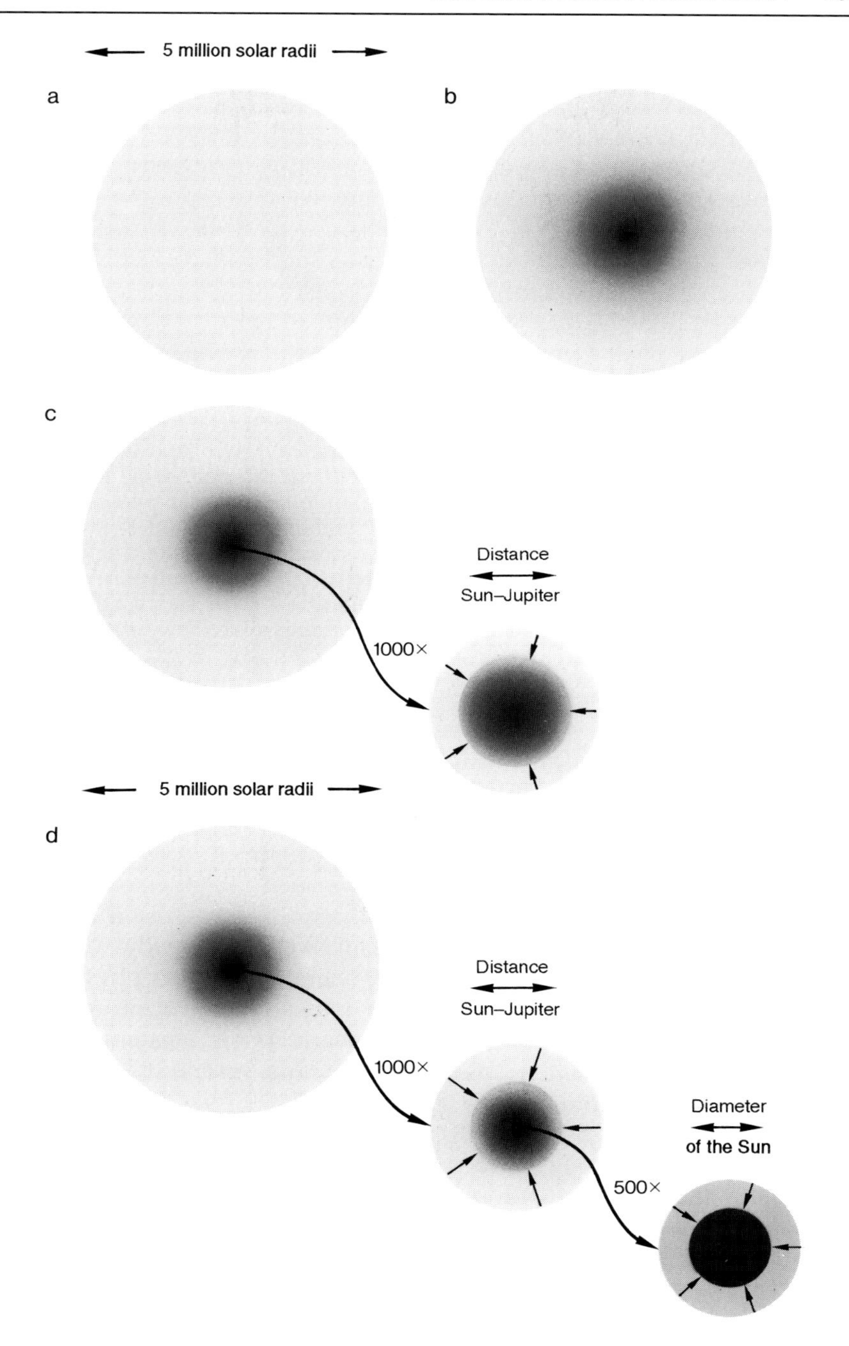
5 million solar radii
a
b
c
Distance
Sun–Jupiter
1000×
5 million solar radii
d
Distance
Sun–Jupiter
1000×
Diameter
of the Sun
500×

Figure 9.3. The interior of the Sun. The left-hand face of the section shows where nuclear processes are producing energy inside the Sun. The right-hand sector shows how this energy reaches the surface. Within the interior it is transported by radiation (wavy arrows), and in the outer layer by convection (indicated by cloud-like curls). The latter motion is responsible for the granulation of the solar photosphere. The lower sector indicates the chemical composition of the solar material. Points show the layers in which the Sun retains the original composition of the cloud from which it formed. Here hydrogen amounts to 70 per cent by weight, and the remainder is mainly helium. Open circles show the region in the centre where, in the past 4.6 thousand million years, the Sun has formed new helium, and thus altered the original chemical composition

snail that moved in a straight line would do the job much quicker. Eventually, however, the photons do work their way up into the outer layers, transporting the energy released by the nuclear processes occurring in the core to the surface. In the outermost layers, however, the relatively cool material is not particularly transparent to photons.

As a result, the energy coming from the interior is transported in the outer layers by a completely different mechanism. We have already encountered this in the shape of turbulence, the deadly enemy of all solar physicists. Hot masses of gas rise and colder ones descend. On Earth, they transfer heat from the hot surface of an asphalt road into the overlying air. This process (which is also known as convection), plays an important part in the heat balance of our atmosphere—and not just above asphalt roads. Convection is also important in the Sun, because in the outermost layers it carries the energy from the interior up to the surface. The energy responsible for every single ray of sunlight that falls upon our skin was carried for a distance of some 200 000 kilometres by convection. To solar observers, convection is nothing new, because it produces the granulation (p. 56). One of the properties of the actual Sun is confirmed by our model. This gives us confidence that it is correct.

The model calculations tell us a lot more. Why should we stop at the present? Let us allow the computer to continue its calculations, and more hydrogen to turn into helium, and let us see what happens to the Sun when all the hydrogen in its core is converted into helium.

For a long time the Sun will show no outward signs that it is using more and more of its reserves of nuclear fuel. Only when it reaches the age of about 12 thousand million years—i.e., in about 7–8 thousand million years—will the Sun have formed a core of helium gas and begin to expand. Simultaneously, its luminosity will rise. Over a period of time that is still measured in thousands of millions of years, the Earth will become warmer. The Sun will become a giant star, which is shown in Figure 9.4 in comparison with its current size. Eventually the Sun will become so large that its surface will reach the orbit of Mercury, the innermost planet. This will be engulfed, as will be Venus some time later. The surface of the Sun will come perilously close to the Earth. It is possible that it will also engulf our planet. While the Sun continues to expand, its luminosity will simultaneously become even greater. Life on Earth will long since have come to an end. But that will not happen for many thousands of millions of years, so we shall be unable to experience this astrophysically interesting time.

After the surface of the Sun has expanded to about the distance of the Earth's orbit, it will begin to shrink. This process will continue, even after it has again reached its current size. Its luminosity will decrease far below its current value. This Sun will become an insignificant star, known as a *white dwarf*, hardly larger than the Earth.

This vision of the future does not just arise from our computer calculations. We observe stars in the sky that are of the same sort as the Sun, but which are more advanced in their life-cycle, and which have expanded (and thus increased in luminosity), as their store of hydrogen became exhausted. We also observe stars that have gone far beyond that stage of their evolution and which are now insignificant white dwarf stars. They have become so dense that a thimble-full of material would amount to about a tonne.

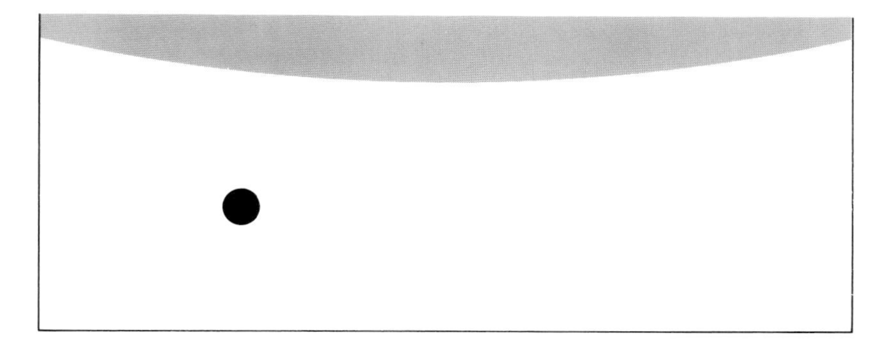

Figure 9.4. Comparative sizes of the present-day Sun (left) and future evolutionary stages. A portion of the giant star that it will become is shown (top), and, hardly visible (right) the size of the eventual white dwarf

Because we can follow the evolutionary history of other stars in our computer by similar methods, and because the results by and large agree with other stars that we actually observe, we have grounds for believing that the picture of the interior of the Sun provided by our computer is the true one, and that the rather gloomy predictions about its, and our, future are correct.

SOLAR RESEARCH UNDERGROUND

One might expect astronomers to be extremely happy with their explanation of the Sun's fusion reactions. The fact that they are not is the fault of the neutrinos that are produced by the nuclear processes inside the Sun. Neutrinos interact with matter so weakly that they stream at the velocity of light more or less unhindered from the Sun's core, where they are produced, and subsequently reach the Earth. If there were such a thing as a neutrino telescope, the Sun would appear as a tiny disk, only about one tenth of the diameter of the Sun at optical wavelengths. Only the central region of the Sun where fusion is occurring would be detectable with a neutrino telescope. There are no such telescopes, however, because neutrinos hardly ever interact with terrestrial matter. It is not possible to construct lenses or mirrors for neutrinos. They are so penetrating that they are not even stopped by the mass of material in the body of the Earth itself. Every second, the Earth encounters about 66 thousand million neutrinos per square metre. They still reach us after sunset. At midnight they come up through the Earth.

Terrestrial material is not, however, completely transparent to neutrinos. Just a few types of atoms are affected by incident neutrinos. The best-known of these is an isotope of chlorine, ^{37}Cl. Very rarely, on odd occasions, it captures a neutrino and, by emitting an electron, turns into an argon atom.

This is the basis of an experiment that has been giving astrophysicists headaches for a long time. Chlorine, in the compound perchloroethylene (C_2Cl_4) is exposed to solar neutrinos in a large tank. The material is a liquid, which is normally mainly used as an industrial solvent, somewhat similar to the better-known carbon tetrachloride. Raymond Davis from the University of Maryland, who developed this experiment, used 390 000 litres of the compound. The same argon atoms may be created, however, not just by solar neutrinos, but also by the protons and other cosmic-ray particles that ceaselessly bombard us from space. To minimize these effects, the equipment was built in an abandoned gold mine about 1500 metres underground, where no cosmic-ray particles would remain.

In addition, to avoid other unwanted reactions from fast neutrons, the tank was surrounded by a thick layer of water. Unfortunately the chlorine atom has the disadvantage that it reacts only with high-energy neutrinos. What energies do solar neutrinos have? Although we may not been able to pass them through a form of terrestrial 'neutrino spectrograph' and spread them out according to wavelength, we can, thanks to our model Sun, determine the Sun's 'neutrino spectrum' theoretically. The chlorine atom reacts only with the high-energy neutrinos that are produced by the decay of boron atoms in a relatively unimportant, secondary reaction chain (see Figure 9.1). They are, however, a vanishingly small fraction of the Sun's overall neutrino radiation. The major neutrino flux produced by the reaction that provides most of the energy (see the first diagram in Figure 1.2) consists, unfortunately, of low-energy neutrinos to which the chlorine atom is blind. I do not intend to discuss the methods by which one detects the argon atoms that have been produced by the solar neutrino flux, nor how one goes about searching for 35 argon atoms in 650 tonnes of perchloroethylene. But I do want to discuss the perplexing results.

A seven-year series of measurements from this experiment, which has now come to an end, is given in Figure 9.5. To have a suitable unit in which to express the results, we use something known as the 'solar neutrino unit' (SNU), which corresponds to the absorption of one neutrino per second in a tank of 10^{36} chlorine atoms. From our model of the Sun it is possible to predict the neutrino flux, and from the properties of the chlorine atom we can derive the argon production rate. The computer models of the Sun indicate a value of 5.6 SNU, far higher than what we obtain. The result of a total of 42 observing runs with the equipment is close to just 1.3 SNU. It is, of course,

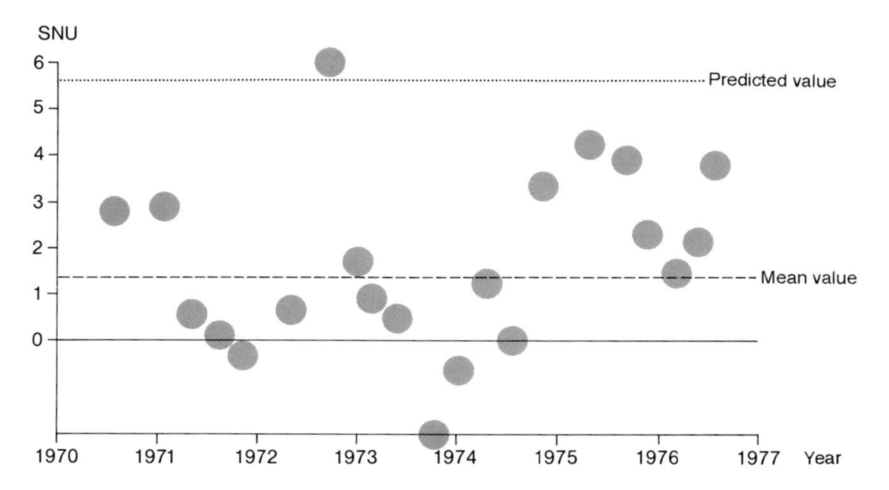

Figure 9.5. Results from 21 test runs of Davis's chlorine experiment. Throughout the period covered, the measurements obtained—with one exception—lie below the value expected from computer models. The mean value of all the measurements is well below the expected level. The discrepancy has not yet been explained. Certain individual values are negative—i.e., fall below 0 SNU—because an average value for unwanted background reactions has been subtracted from the measurements. Because the true number of spurious reactions varies statistically, the solar neutrino rate occasionally assumes a negative value

only possible to give an approximate value, because it is, of necessity, subject to considerable errors. It is, however, certainly far from agreeing with the expected value of 5.6 SNU.

When the first results of Raymond Davis's experiment started to emerge in the 1970s, astronomers were not very concerned. Computer models of other stars agreed with many observed properties. Even the solar model accurately reproduced nearly all the Sun's properties—except the neutrinos. The solar models were probably correct, and there was something wrong with the experiment. But no one was able to point out any errors to Davis. But then people began to say: He is right, there are significantly fewer neutrinos emerging from the Sun than expected.

The news emerged from a Japanese zinc mine near the town of Kamioka, 300 km west of Tokyo. There, about 2140 tonnes of water have been monitored by light detectors since 1986. In fact, this is not an astronomical experiment. What is being investigated in the mine, 1000 metres underground, is whether the proton has an infinite lifetime, or decays. Modern theories suggest that it has a very long lifetime, but leave open the possibility that, on average, this may be approximately 10^{29} years. To our minds, 10^{29} years may seem a very long time for a proton to survive, but 2140 tonnes of water contain about 10^{30} protons. On average, therefore, we should observe one proton decay about once every month. Because this is not observed, it is believed that the lifetime of a proton must be at least 10^{32} years.

The apparatus in the Kamioka mine does not just detect the spontaneous decay of protons, however. Energetic neutrinos from the Sun, emitted by the ^{8}Be reaction shown in Figure 9.1, may occasionally collide violently with an electron. The high-velocity electron itself creates a flash of light. A battery of detectors watch the 2140 tonnes of water, day in, day out, for these flashes of light. The number of solar neutrinos was determined over a period of 450 days between January 1987 and May 1988. Only about half of the number predicted by solar models were found. Raymond Davis' experiment was fully confirmed. Once again, solar neutrinos are a thorn in astronomers' flesh. The Sun is unique in that among all the nuclear reactors in the Solar System, it is the only one that disappoints us by exhibiting a lower level of particle radiation than expected.

What is wrong with our solar models? At first glance, the discrepancy does not appear to be too bad. The neutrinos that do not show up in the chlorine experiment are produced by a secondary reaction that is unimportant in the Sun's energy-production process. A slightly lower central temperature for the Sun would bring everything into complete agreement, without, however, altering the Sun's other observed properties to any great extent. For 20 years astronomers have been searching in vain for errors in their solar models, and hoping that someone would find a physical basis for why the Sun's central temperature should be slightly lower.

Only when the models are artificially altered can the computer Sun be brought into agreement with the two experiments. We can, for example, arbitrarily assume that the Sun is constantly mixed, and that the helium produced in the core is distributed throughout a much larger volume—perhaps even throughout the whole Sun. The central temperature would then be slightly lower, just enough to agree with the neutrino flux measured by the chlorine experiment. Against this, however, is the fact that we do not know what might create constant mixing of the solar interior.

There is another way out that I want to mention. It was proposed by the Californian astrophysicist William Fowler. We need first to conduct a thought experiment. Let us assume that, abruptly, the temperature at the Sun's core were to drop so low that the nuclear reactions cease. What effects would we detect? Immediately, i.e., in eight minutes (which is the travel time of the neutrinos, moving at the speed of light), the solar-neutrino flux would drop. But it would take about 20 million years for the Sun's luminosity to react to this change. Immediately—like an emergency power supply during a power cut—the collapse mechanism proposed by Helmholtz that we discussed in Chapter 1 would take over supplying the Sun's energy. As we saw, that would suffice to maintain the Sun's luminosity for 20 million years.

So we could therefore explain the measured neutrino deficit as follows: For whatever reason, the temperature in the central regions of the Sun varies. At present we are probably in a phase of relatively low central temperatures. As a result, less solar energy is being produced, and therefore fewer of the temperature-sensitive high-energy neutrinos are emitted by the Sun. At present, therefore, not all of the Sun's energy originates in nuclear reactions, but some derives from the mechanism suggested by Helmholtz. The Sun is contracting so slowly that even over a period of thousands of years we would observe nothing. After some 10 million years we would probably again have a hot core and high neutrino-production rates. More energy would be released than radiated away by the Sun. The excess energy would be consumed in expanding the Sun from the smaller radius that resulted when less energy was produced to its original size. So we could explain the neutrino experiments by saying that temperature variations occur within the Sun, and that these are accompanied by variations in the neutrino flux. We happen to be in a period when the neutrino flux is low.

So far, however, all attempts to show that solar models must undergo temperature variations have failed. Years ago, John Bahcall, an astrophysicist at the Institute of Advanced Studies in Princeton said to me, after a year-long investigation of solar models for potential instabilities, 'The Sun is as stable as a rock.'

MINING SOLAR RADIATION FROM ORE

Anyone driving across the Rocky Mountains in Colorado has the opportunity to stop at Clear Creek and, where the small stream runs close to the road, to pan for gold. There are pans lying around to be used, as well as a pick-axe, and a box in which prospectors may leave, at their own discretion, a contribution to the owner of the land for the gold that they have found. Many years ago I was unable to resist seeing what I could find. I kept the mixture of gold and sand for a very long time. Between the grains of sand there were impressive clumps of gold—although you needed a high-power microscope to see them. Alongside the remnants of some long-forgotten gold-digger's lucky strike, Clear Creek probably still holds a different type of treasure. This is because this region is rich is an otherwise fairly rare mineral, *molybdenum glance* (or molybdenite).

The heavy metal molybdenum resembles lead. It is used in various alloys, one of which is the particularly strong, corrosion-resistant molybdenum steel. The molybdenum is obtained from ores that contain molybdenum-bearing minerals, above all, molybdenite. These ores are found in Colorado in the USA, but also in Australia and Norway. Like nearly all the chemical elements, naturally occurring molybdenum consists of several isotopes, among which is molybdenum-98, ^{98}Mo. Like the chlorine ^{37}Cl and the gallium ^{71}Ga isotopes, this occasionally reacts with a neutrino. When a neutrino collides with an atom of ^{98}Mo, the molybdenum atom may occasionally become an atom of technetium, more specifically ^{98}Tc. In doing so, an electron is emitted.

In principle, the processes are similar to those that occur in the chlorine experiment. There a chlorine atom occasionally becomes an argon atom. The difference, however, is that the argon in the tank of chlorine itself decays after 35 days, whereas the technetium atoms exist for six million years! If the Sun has not changed significantly over that period of time, a balance must have been established in the molybdenum present on Earth, in which as many technetium atoms are formed by interactions with solar neutrinos every second as are destroyed through natural radioactive decay. So if we can determine the abundance of ^{98}Tc atoms, we also know how many decay every second. The same number of technetium atoms are produced by solar neutrinos in the same period. This allows us to determine the neutrino flux from the Sun from the quantity of ^{98}Tc on Earth.

Unfortunately, the molybdenum atoms, once again, provide information solely about the high-energy neutrinos. Technetium abundances are therefore evidence for neutrinos that are also detectable by the chlorine experiment.

To count the few ^{98}Tc atoms in molybdenum-bearing ore reliably, we have to begin with 2600 tonnes of ore. From this 13 tonnes of molybdenite are obtained. One may expect these 13 tonnes to contain about 10 million atoms of ^{98}Tc.

In contrast to the chlorine experiment, where we have to count argon atoms that are a few weeks old, we now need to count technetium atoms that have an average age of several million years. The number present will reveal something about the Sun's neutrino radiation in the last few million years. Because of its technetium atoms, molybdenite has a very long memory.

It may reveal that the Sun has behaved differently over the past few million years from what we believed previously.

THE EXPERIMENT BENEATH THE ABRUZZI MOUNTAINS

Is there a crisis in solar physics? So far we have consoled ourselves by saying that the boron neutrinos, which are the only ones that can affect the chlorine atoms, arose from a secondary reaction that is insignificant in terms of the solar energy production. Neutrinos from the all-important reactions have not been measured, and they are the ones that actually matter.

This situation will not last long, however, because the hour of truth for astrophysicists is approaching. Recently, a search for the low-energy solar neutrinos—the ones that arise in the proton—proton chain (Figure 1.2, top)—has begun. The possibility of doing so is offered by the gallium atom. A collaborative, international project, known as Gallex, has been set up. The scientific director is the physicist Till Kirsten from the Max Planck Institute for Nuclear Physics in Heidelberg. In a cavern hollowed out of the rock alongside the Gran Sasso road tunnel beneath the Abruzzi mountains in Italy, 1200 metres beneath the surface, 30 tonnes of gallium are exposed to solar neutrinos. Figure 9.6 shows a diagram of the underground laboratory.

Gallium, a white metal that melts even at human body temperature, consists of 40 per cent ^{71}Ga, the neutrino-sensitive isotope. When one of the gallium atoms reacts with a neutrino, it turns into a germanium isotope, ^{71}Ge.* The ^{71}Ge decays back into gallium with a half-life of 11.4 days. It was estimated that inside the tank of gallium under the Abruzzi mountains, one gallium atom per day would be converted into germanium. In a month, about 20 germanium atoms should be produced. It is possible to find these 20 germanium atoms in 30 tonnes of gallium. By a series of intermediate chemical stages, the newly created germanium is linked with hydrogen to form a gas, somewhat similar to methane. When this gas is examined, the ^{71}Ge atoms are detectable, thanks to their radioactivity. The experiment began in 1990, and should enable a firm conclusion to be drawn after a period of about four years as to how many neutrinos, produced by the Sun's principal nuclear process, are actually reaching the Earth.

The neutrino experiments offered the possibility of looking inside a star. When the number of boron neutrinos was determined, the results were embarrassing: something in the deep interior of the Sun, where the fusion reactions are taking place, was not as we expected. The current results from

* Given the names of the elements involved, gallium and germanium, this experiment seems to call for Franco-German cooperation. In fact, physicists from Nice and Saclay in France, and from Heidelberg, Karlsruhe and Munich in Germany are working alongside Italian colleagues from Milan and Rome. In addition, workers from the USA and Israel are also involved in the task of spotting each day's germanium atom.

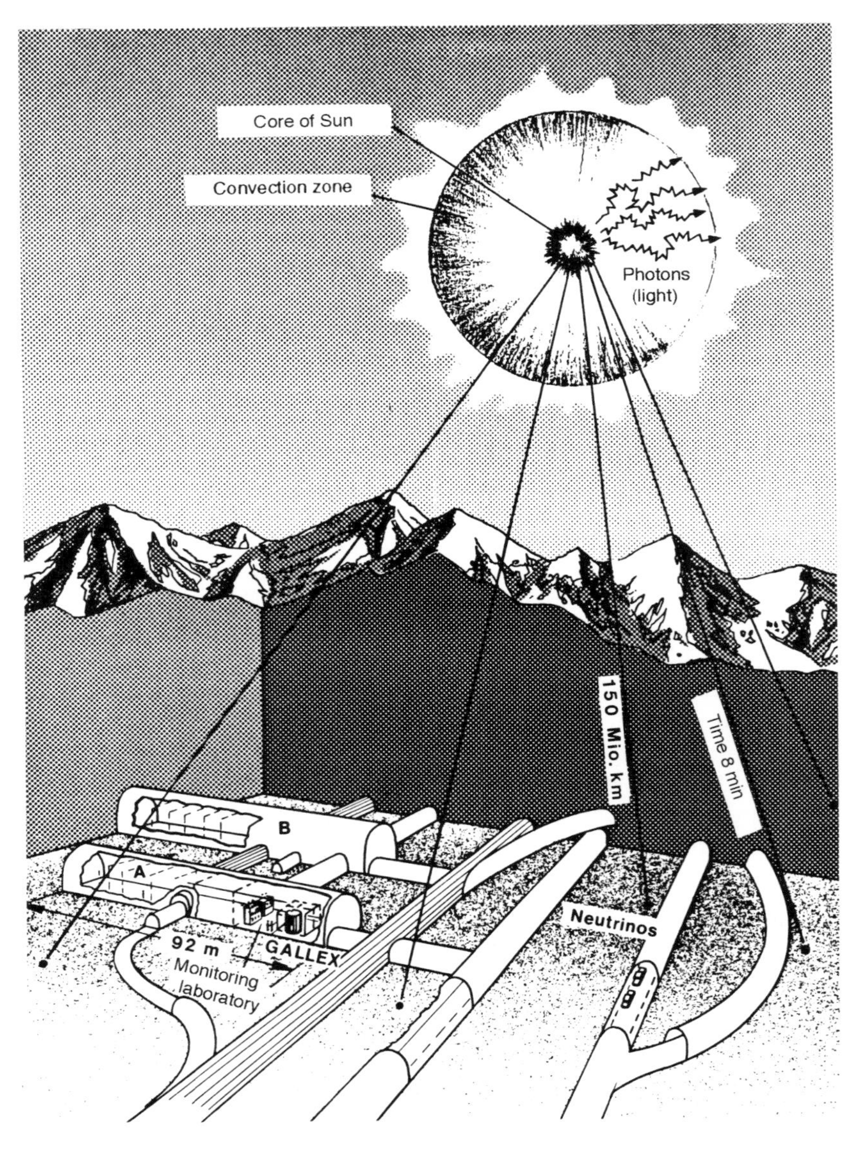

Figure 9.6. A sketch of the laboratory in which the Gallex neutrino experiment began monitoring solar neutrinos early in 1990. The neutrinos reach the Earth unhindered from the centre of the Sun, penetrate the mountains of the Gran Sasso massif, and encounter a tank full of a gallium compound, located in a cavern carved from the rock alongside the road tunnel

Gallex allow astrophysicists to sleep somewhat easier. The low-energy neutrinos are appearing at the expected rate—regardless of what is effecting the high-energy boron neutrinos. Recently, it seemed as if the Sun had yet another surprise for us.

WAS THE SUN ONCE LARGER?

Theophilus Shelton cannot be found in reference books. He would have been astonished to learn that, nearly 300 years later, astronomers would like to know exactly where his house was situated. He would have been still more surprised if he were told that the reason for this was that people wanted to test whether the Sun had become smaller over the centuries since his death.

It all began in 1987 with an announcement by several astronomers from the Paris Observatory that astonished their colleagues around the world. The Parisian team had studied data about solar eclipses occurring between 1666 and 1719 that had been acquired by Jean Picard—whom we met on p. 2—and other French astronomers. Following their analysis, the team came to the conclusion that the Sun must have been larger than it was previously and than it has been since. The old measurements could best be explained if the Sun were 2000 kilometres larger than it is today. This may be only 0.3 per cent, but it worried astronomers that such a fundamental property as the radius of the Sun should vary for no apparent reason. The aforementioned period largely coincided with the Maunder Minimum (p. 29). Could this be a clue to the anomalous solar activity at that time? Solar researchers were soon reassured, however, thanks in no small measure to Theophilus Shelton from the English village of Darrington in West Yorkshire.

In May 1715 the 300-kilometre-wide track of a total solar eclipse lay across the southern half of the British mainland. Edmond Halley (1656–1742), who is best known for the comet named after him, appealed to the public to observe the event and to send reports to him, even if observations were only made with the naked eye. From the details submitted to him he was able to determine very accurately the width of the path of totality, i.e., the track on the ground where the Moon completely covered the Sun. On the southeastern edge, William Tempest from Cranbrook in Kent recorded that he had seen the Sun completely covered for just a brief instant. On the northwestern edge, Theophilus Shelton reported that the portion of the Sun remaining uncovered at maximum eclipse was 'the size and brightness of Mars'. Yet at the nearby village of Badsworth the solar corona was seen. That point was therefore within the path of totality. Outside that track the eclipse was a partial one

Alerted by the paper from Paris, a group of astronomers from the Royal Greenwich Observatory re-examined the reports collected by Halley. Above all, they succeeded in determining exactly where Shelton's house was situated. If the Sun had been as large as the group from Paris maintained, the path of totality would have been eight kilometres narrower. Shelton would not have seen the Sun as a point from his house, neither would Tempest have been able to see it fully covered for about one second. As seen from Badsworth, the corona would never have been visible against the sky brightly illuminated by the uncovered portion of the solar disk.

Consequently, the Sun was the same size in 1716 as it is today. As far as this was concerned, the astronomers' world was once more secure. Only the neutrino problem remained. Neutrinos offered, for the first time, a chance to observe what was happening deep inside the Sun. Subsequently, a second method of studying the behaviour of the solar interior has been devised.

10

OSCILLATIONS OF THE SUN

> A full wineglass has a deeper ring when it is struck than an empty, or half-full one. A flute blown with hydrogen has a higher range of notes than in ordinary use. . . . When it was discovered ten years ago that the oscillations of the Sun, which had been known for a long time, could be ascribed to . . . harmonics, it suggested that the same physical principles that were the basis of the experiments with a wineglass or a flute, could also be applied to the gigantic resonator that is the Sun.
>
> Franz-Ludwig Deubner (1985)

In recent decades a new method of learning something about the invisible interior of the Sun has been discovered. All that we need to do to find out what is happening is to study the surface very carefully.

The primary cause of the motion of material at the surface of the Sun is the circulation in the granulation. The columns of rising and falling gas have diameters of about 1500 kilometres, which is about 1 per cent of the solar diameter. The Doppler effect provides us with a means of determining the velocities involved, which are about one kilometre per second. Within minutes the motions fade away, to be replaced by new granules. Apart from the granulation there is also the supergranulation, which is slower, and thus larger and longer-lasting.

Amidst this chaotic swirl of different sized bubbles of material, no one can predict when one of them will break up, or in which directions its parts will subsequently disperse. If all this motion towards and away from us, and the flow, now towards the left and now towards the right, is seen in time-lapse photography, the seething surface of the Sun reminds us of the chaotic, unpredictable motion of boiling water. Imagine looking down on a point on the surface of the water. At one moment it is being lifted upwards towards us, at the next it is moving sideways. Sooner or later it will disappear back into the depths again. The motion of the granulation's bubbles of gas in the outermost layers of the Sun is similar, only faster.

Among all this incomprehensible muddle, people have found various regularities that no one had previously suspected.

THE SUN'S FIVE-MINUTE OSCILLATION

The new era of solar physics began in 1960. At the California Institute of Technology, a team under Robert Leighton (once again) was investigating the solar granulation. The aim was to follow the evolution of a single granulation cell from its birth to its final decay. To do this, it was essential to be able to measure the velocity of a specific, fixed point on the Sun's disk (using the Doppler effect). The team knew approximately what they could expect.

We can illustrate this by considering a pot of boiling water. Let us assume that we want to determine the vertical velocity of a particular point on the boiling surface by using a float connected to a rack and pinion mechanism that moves a pen, which leaves its trace on a steadily moving strip of paper. The arrangement is shown schematically in Figure 10.1a. The recorded curve shows no sign of regularity. If we were able to increase the frequency of the motion until it could be detected by ear, we would hear a rushing sound ('white noise') like that from a badly tuned radio or television set. No individual tone would be detectable. But Leighton and his colleagues found that the Sun did not just emit white noise.

They were looking at an area of the surface that was large enough to include a considerable number of granules, which were therefore both rising and sinking. Their velocities could therefore be expected to more or less cancel out. Yet there remained a residual velocity variation that was amazingly regular. While the granules in the area being investigated were rising and sinking, the whole surface of the Sun was simultaneously doing the same. The velocities, at about 0.5 kilometres per second, were less than those of the granulation, and they reversed rhythmically with a period of about 296 seconds, i.e., just under five minutes. One of the first measurements is shown in Figure 10.2. The oscillations were not always equally prominent. Sometimes their amplitudes were greater, sometimes smaller. We can illustrate the observed oscillations by again considering our pot of water.

Let us now assume that the pot is hung from a spring and that, before the measurement began, we raised the pot slightly, and then released it. The boiling water would rise and fall with the pot. Our equipment would record a regular oscillation. The result is shown in Figure 10.1b. If the motion were now rendered audible, a regular variation, a specific note, would be superimposed on the white noise. It was this note that Leighton's team discovered in the Sun.

Is the Sun's surface oscillating? It has been known for a long time that certain stars pulsate regularly. The most famous are the Delta-Cephei (or 'Cepheid') stars, with expand and contract with a period of several days.

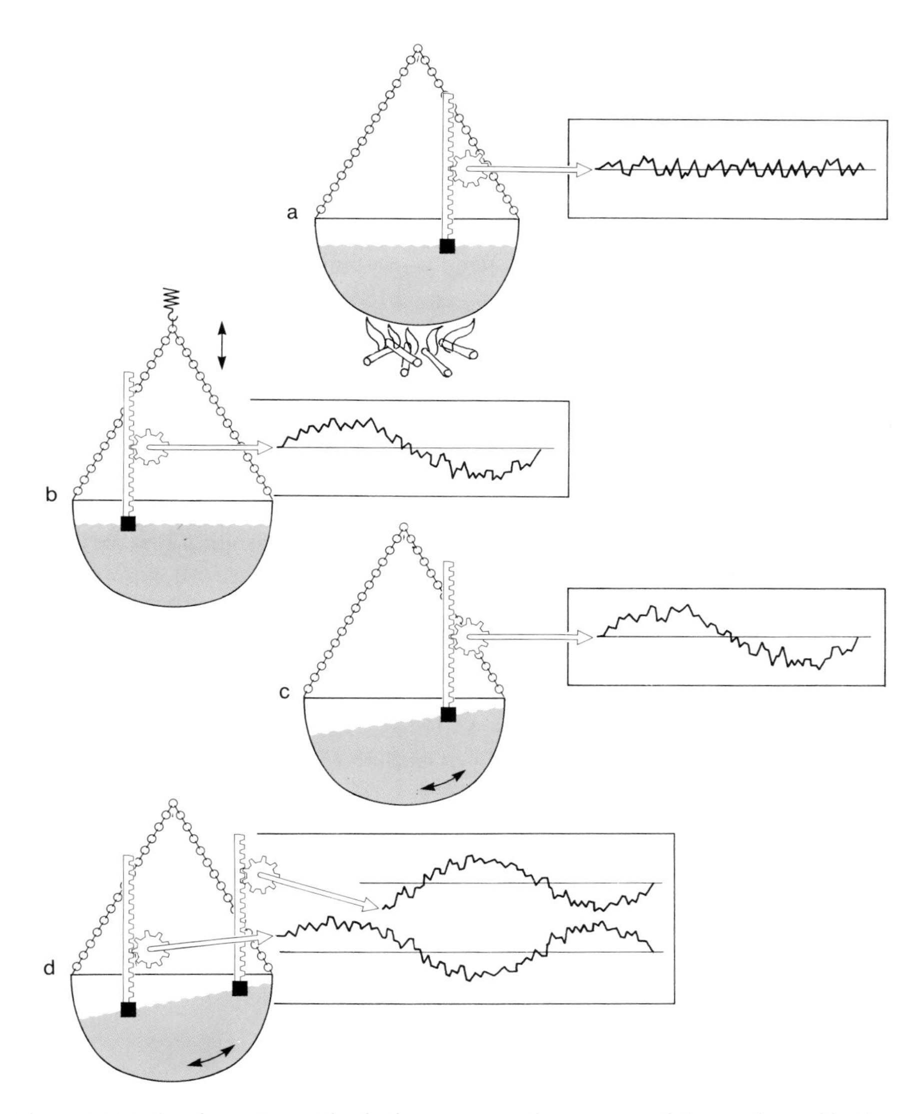

Figure 10.1. A schematic method of measuring the motion of the surface of boiling water. (a) The boiling water shows an irregular vertical motion. (b) If the container is hung by a spring and moves up and down, a regular wave is superimposed on the irregular motion. (c) If the liquid is set in motion by a sideways push, it is not possible to tell, with a single measurement, whether the surface of the water is moving upwards and downwards as a whole (as in b), or whether it is oscillating backwards and forwards. (d) Only two measurements allow one to determine the difference

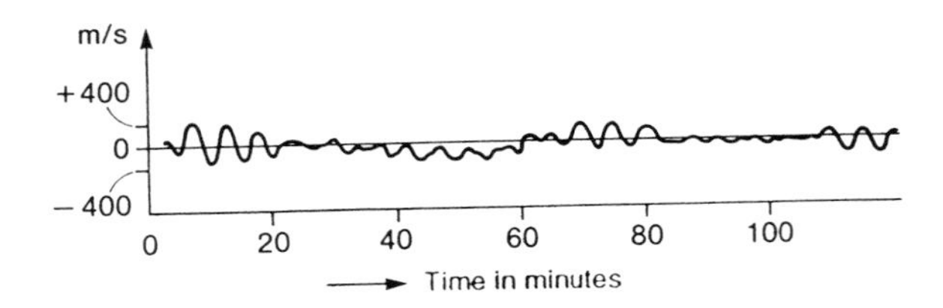

Figure 10.2. The motion of a specific point on the solar surface towards and away from the observer always exhibits a five-minute rhythm

Could the Sun perhaps be pulsating with a period of five minutes? Does its diameter increase for about $2\frac{1}{2}$ minutes, and then decrease during the following $2\frac{1}{2}$ minutes?

Let us look again at our pot of water. It is not essential to suspend it from a spring if we want to introduce a 'note' into the recording. All we need to do is to give the pot a jolt before we start the measurement. Then the water begins to oscillate (Figure 10.1c). Provided we are not taking our measurements in the exact centre of the vessel, we obtain a curve that is identical to that obtained when the pot is suspended by a spring. The recordings are identical in both Figure 10.1b and 10.1c.

In addition, we need to measure two points inside the vessel, roughly as shown in Figure 10.1d, where linkages simultaneously register the motion of the water in two different places. If we let the vessel oscillate up and down, both linkages would record 'water up' and 'water down' at the same times. If the water is fluctuating as shown in Figure 10.1d, then one recording channel would record 'water up' while the other registers 'water down'.

Simultaneous measurements at numerous points on the Sun have shown that it does not oscillate like the pot suspended by a spring, but rather like the water in Figure 10.1d. Everything is far more complicated, however. I must now ask the reader's indulgence if we first discuss the behaviour of vibrating bodies. Hundreds of years before anyone suspected that the Sun vibrated, oscillations of terrestrial bodies had been studied extensively.

VIBRATING STRINGS

It is said that Galileo Galilei first became interested in the laws governing a pendulum when, as a student, he compared the duration of the swings of a chandelier with his heartbeat. It is not just pendulums that swing regularly, all musical instruments produce vibrations in the air—the bowed string of a violin is just one example. For simplicity, let us imagine that our string is not part of a musical instrument, but is stretched between two fixed points, as shown schematically in Figure 10.3a.

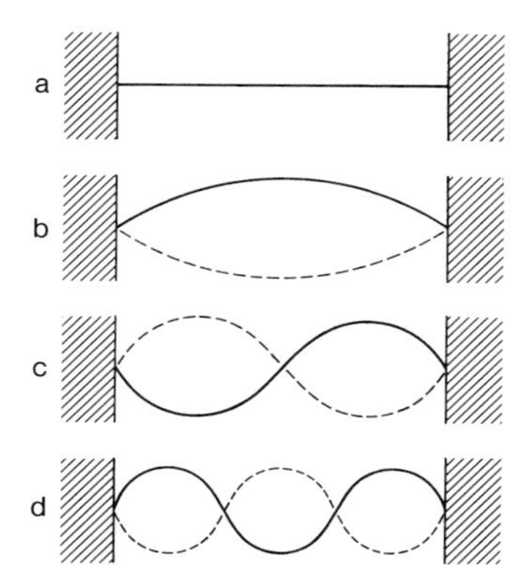

Figure 10.3. If a string that is stretched between two fixed points (a) is plucked in the centre, then it vibrates with the maximum amplitude in the centre (b). It is also possible to pluck it using both hands in such a way that a node forms at the centre, accompanied by two antinodes, one on each side (c). In (d) the string is vibrating with two nodes and three antinodes. The string's excursions are greatly exaggerated

If we pull the string towards us in the centre and suddenly release it, the vibration is as shown in Figure 10.3b. Naturally at the ends, where the string is fixed, there is no motion. The movement is greatest in the centre. This point is known as the *antinode.* When a vibration has just *one* antinode it is termed the *fundamental vibration* of the string. It produces the deepest note. The pitch of a note is also influenced by the tension in the string, as anyone will know who has watched while violinists tuned their instruments. But the way in which a string of a given length and tension vibrates also determines the note that is produced.

If, instead of plucking the string in the centre, we use both hands to pull one half of it upwards and the other half downwards, and then release both hands simultaneously, the string vibrates in a different manner. We hear a higher note, and closer inspection shows two antinodes, where the displacement is in opposite directions (Figure 10.3c). When the string is at its maximum displacement upwards on one side, it is at its maximum displacement downwards on the other. The two halves are said to be in opposite *phase.* In the centre the string is essentially stationary, and this point is known as the *node* of the vibration. A vibration of a string with just one node (and two antinodes) is known as the *first overtone.* Violinists deliberately use this sort of vibration, or resonance mode, which is known as the 'flageolet tone'. They produce it by lightly touching the exact centre of the string during bowing.

If the string is plucked with three hands, the result is the second overtone (shown in Figure 10.3d), which has two nodes. The two outer portions of the string vibrate with precisely the same phase. The centre vibrates in the opposite phase. In principle, more hands can be used to obtain still higher overtones, with more nodes and antinodes.

I have considerably simplified the situation. We would never actually succeed in producing such vibrations. If one tries to excite the fundamental note of a string with just one hand, it will actually produce a range of overtones, all superimposed on the fundamental. When violinists set the strings of their instruments vibrating with their bows, they produce a whole series of overtones, which, together with the fundamental, create the characteristic tone of violins. We can recognise this by ear alone. We can tell whether a specific note has been produced by a violin, a trumpet, or a flute. Particular overtones are enhanced in different instruments. This is why we are able to hear which instrument produces a particular note, because the mixture of overtones gives an instrument its characteristic sound.

The vibrations that we have discussed are known as *standing waves* (or resonances). Nodes and antinodes remain at specific points on the string. A completely different situation arises with *travelling waves*. This is the sort that we see on the surface of the water when we throw a stone into a pond. A simple experiment shows how standing waves may arise from travelling waves.

A VIBRATING ROPE

Instead of our taut string, we now need to imagine holding up a piece of rope and allowing the other end to hang freely. The nature of the material does not really matter, but only the way in which it is suspended. Our string was taut, which was what made it capable of vibrating. A piece of rope hanging down is kept taut only by its own weight. The situation is not very different from the string, however, only in so far as one end is completely free to move. Motion of waves is affected differently when the ends are free than when they are fixed.

Imagine that we briefly move the upper end sideways. The displacement moves downward (Figure 10.4, left). We have a travelling wave. At the bottom it is reflected by the free end, so it then travels back up the rope. Let us carry on, and move the upper end of the rope back and forth rhythmically. Then, we can try to alter the rhythm. With a bit of skill we will manage to produce standing waves. The waves travelling downwards and those reflected upwards by the free end will combine to give a set of stationary waves, with nodes and antinodes (Figure 10.4, right). If we shake the rope even faster, we can produce even more nodes, thus causing the rope to vibrate in higher overtones. We see from this that opposing wave-trains of the same frequency may produce standing waves.

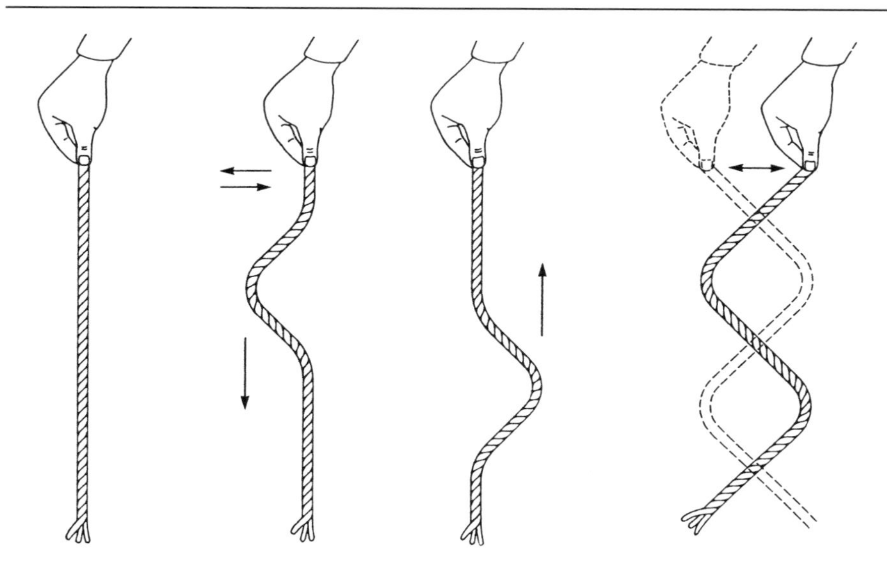

Figure 10.4. If you move the hand holding a hanging piece of rope sideways, a wave travels downwards and then is reflected back. If you shake the upper end rhythmically backwards and forwards horizontally, a standing wave is produced. There are points that are stationary, and these are the nodes of the vibration. (The displacements are greatly exaggerated here)

WAVES IN A BOWL OF SOUP

Strings and ropes are one-dimensional objects, in which waves may propagate in just *one* direction, either forwards or backwards. A wave spreading across the surface of a pond has two available dimensions. In addition to travelling waves, standing waves may also form on a water surface.

We can see this for ourselves with a simple experiment. Take a soup-bowl full of water and move it rhythmically backwards and forwards on its plate. If we get the timing right, 'crests' and 'troughs' always form at the same points of the surface, while at other points the water does not slosh around at all.

The oscillating water now shows *nodal lines*. As waves produced by the rhythmic movement of the bowl travel from the edge to the centre, they combine with waves produced by the opposite side of the bowl to give rise to oscillations that are 'fixed'—standing waves. The same points on the surface always oscillate, and the same points always remain stationary. The pattern of waves that results no longer moves. The waves continuously reflected from the sides of the bowl combine to produce a standing-wave pattern.

It is not just the surface of liquids that may be set into regular standing waves. Solid bodies may vibrate in a similar manner.

THE VIBRATIONS OF A GLASS PLATE

Experimental techniques in physics have made great strides in the last 100 years. If anyone nowadays wants to open up a new field of research, expensive equipment is required with which to produce a plasma (say) or to accelerate electrically charged particles to velocities close to that of light. Two hundred years ago, a glass plate, a violin bow, and some powdered sulphur made the scientific world sit up and take notice.

In 1787, Ernst Florens Friedrich Chladni (1756–1827), a lecturer at the University of Wittenberg, took a glass plate, held it at one corner with a clamp, and scattered powdered sulphur on the surface. He then took a violin bow and stroked one of the plate's free edges. Like a string, the plate began to vibrate. The vibration caused the sulphur particles to be shaken away from

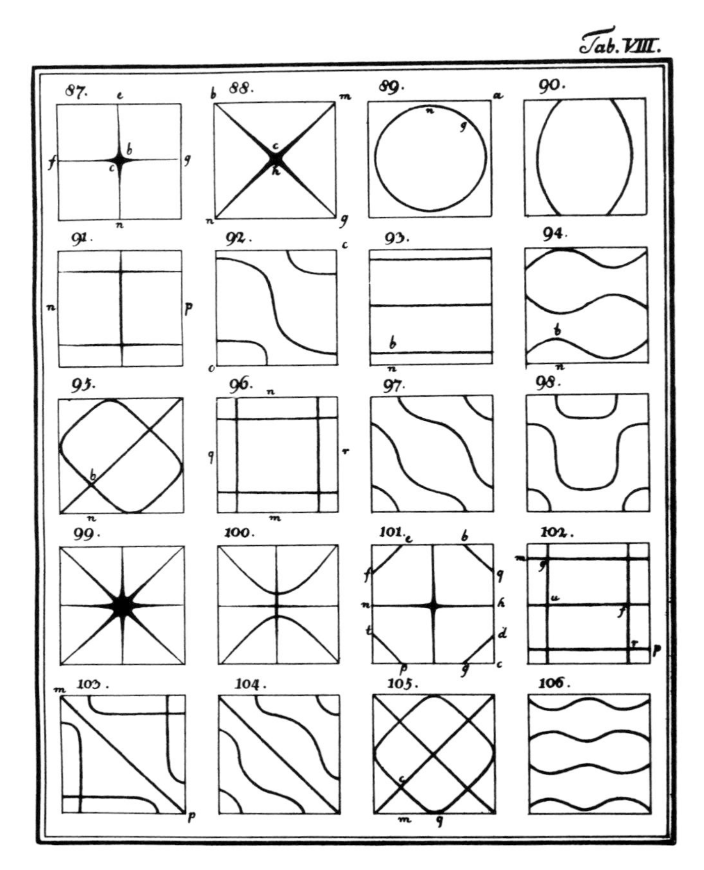

Figure 10.5. The resonant modes of a glass plate excited by a violin bow may take the most varied forms, depending on which point of the plate is clamped, and where the bow is applied (Deutsches Museum, Munich)

the points where the plate was in motion until they reached stationary points, i.e., the vibration's nodal lines. They remained there, revealing the nodal lines, which formed regular patterns (Figure 10.5), now known as Chladni figures. Different figures corresponded to different tones of the vibrating plate. What caused these figures? From the point excited by the bow, waves of a specific frequency ran out over the plate. They were reflected by the edges of the plate, and combined with waves coming directly from the bow. A point on the plate fell on one of the nodal lines if all the waves reaching it—whether direct or as echoes from the edge of the plate—all cancelled out. At any point on the nodal lines, the combined waves reaching it produced two forces (one attempting to raise the surface, and one attempting to lower it), that were precisely equal and opposite. As a result, the point was stationary. The sum total of all the superimposed waves gave the observed pattern of nodal lines.

At first sight, it may appear surprising when we say that Chladni's figures were produced by waves that ran across the plate at a specific rate, because it is quite obvious that the sulphur particles clustered along the nodal lines remain perfectly stationary on the plate. This is irrelevant. We are not seeing all the individual travelling waves that affect a particular point. What we are seeing is the pattern of standing waves that they form in combination, just like the two opposing wave-trains seen in Figure 10.4.

VIBRATING SPHERES

We have seen that the standing waves in a one-dimensional object, such as a rope, form nodal *points*, but that two-dimensional objects, like Chladni's plate, have nodal *lines*. What about three-dimensional objects? Because our aim is to understand vibrations in the Sun, we shall consider just spheres of gas that are held together by their own gravity.

The various resonant patterns that may, in principle, affect spheres of gas may be investigated by using a computer. So many have been found that it is difficult to distinguish between them. Figure 10.6 shows a few of the simplest ways (or *modes*) in which the surface of a sphere may vibrate. When

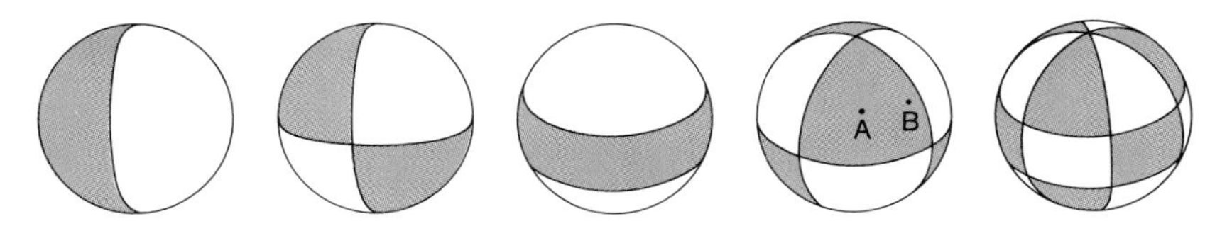

Figure 10.6. Various modes of vibration of a sphere. The white and grey areas, which are separated by the black nodal lines, indicate regions of the sphere's surface that at any one time are moving outwards and inwards, respectively. As explained in the text, the degrees of oscillation, reading from left to right, are: one, two, two, three, and five

the points within the white areas are moving outwards, those in the grey areas are moving inwards. At the boundaries between the white and grey areas the sphere is stationary. These boundary lines are the nodal lines. They form 'great circles' (or 'meridians') and 'small circles', similar to the lines of longitude and latitude that are familiar to us from geography. The 'meridians' intersect at the vibrational 'poles'. The great circle equally distant from both 'poles' is the vibrational 'equator'. The total number of nodal lines, i.e., both great and small circles, gives the *degree* of oscillation. The larger the number of nodal lines, the higher the degree.

When it comes to a sphere, however, we must consider more than just the surface. We must also take the interior into account. This is shown in Figure 10.7. We can see that throughout the sphere, regions that are rising and others that are falling exist simultaneously. These regions are separated by surfaces, which are therefore the nodal *surfaces* of the oscillation. We find that nodal lines occur where the nodal surfaces intersect the solar surface.

Figure 10.7. A section of a sphere of gas pulsating in a particular vibrational mode. The cut-away section reveals that the nodal *lines* on the surface are actually determined by the nodal *surfaces* within the interior. Where a nodal surface intersects the outer surface of the sphere it gives rise to a nodal line. There are also other nodal surfaces, however, that are concentric to the centre of the sphere and never intersect the surface. They cannot be detected by examination of the exterior. Only more detailed study of the degrees and frequencies of oscillation enables us to determine the number of spherical nodal surfaces hidden below the sphere's exterior

GRAVITY WAVES AND SOUND WAVES

In principle, the oscillations in our sphere of gas may be ascribed to two types of wave. They may, for example, be like waves on water, or they may be a different form of wave that is not found on the surface of a lake. Our sphere consists of gas, which may not only be easily moved backwards and forwards like a liquid, but may also be compressed like the Earth's atmosphere. It is not only when we pump up a car tyre that air is compressed. Even the slightest vibration of our vocal cords, or the faintest rustle of tissue paper causes a slight compression of the air. It expands again immediately, but overshoots the mark. It momentarily becomes less dense than previously, and thus includes a slightly larger volume of space than before. This can only happen if neighbouring regions of gas are compressed. They react in exactly the same way, and compress their neighbours in turn. Each momentary compression of the air produced by our vocal cords is propagated outwards. The result is a sound wave that travels though the air at the speed of sound.

The Sun's vibrations may stem from sound waves that criss-cross the interior from every part of the surface. Let us therefore imagine that sound waves of a specific frequency run through our sphere of gas. When they reach the surface, they are reflected back into the interior, like the wave from the end of our piece of rope. The waves moving inwards encounter other waves moving outwards and combine to give rise to a pattern of standing waves.

Sound waves are not the only ones that may cause the Sun to vibrate. The oscillations of the solar surface may stem from waves that are similar to those on the surface of water in a lake. They are fundamentally different from sound waves. At the crests of waves on the surface of water, the water is lifted vertically. The Earth's gravity and surface tension pull it back, causing it to fall below its rest level. But then water pressure forces it back upwards again. It overshoots its rest position, and gravity and surface tension drag it down again.* Because of the important role played by gravity, waves like those on a lake are also known as *gravity waves.*

With sound waves gravity plays no part. Instead it is gas pressure that, once the gas has been compressed, causes it to expand again and overshoot its rest position, resulting in neighbouring bodies of gas being compressed, which in turn affect other bodies, and so on. Any compression propagates through the gas at the speed of sound. Irrespective of whether it is sound or gravity waves that are rhythmically affecting points on the surface of the Sun, they create a regular pattern.

---

* We are all familiar with surface tension in the form of soap bubbles. It tends to reduce the surface area as much as possible. This is why soap bubbles and droplets of mercury are spherical, because a sphere has the smallest surface area for a given volume. When waves form on the surface of a lake, the total surface area increases with every wave-crest and trough. As well as gravity and water pressure, surface tension also tends to reduce these irregularities in the surface. Surface tension is stronger than gravity only at wavelengths below two centimetres. There is no surface tension with gases.

After Leighton's team had discovered the Sun's five-minute oscillation, the question was whether the Sun exhibited a regular pattern of vibration, or not. The Sun's resonances are not so easy to detect as those on Chladni's glass plates. This is not solely because we are unable to scatter sulphur powder over it.

Everything would be much simpler if the Sun vibrated in a single mode, such as the one shown on the extreme right of Figure 10.6. In that case the degree of oscillation is five, because there are five nodal lines: three great circles and two small circles. Along any radial line down into the interior there are probably a certain number of additional nodes. In a specific oscillation mode, the Sun vibrates at a frequency that corresponds precisely to this pattern of nodes. We shall see later that the Sun does not exhibit such a 'pure' vibration of nodes and antinodes. Provisionally, however, let us see what we would observe if the Sun did have such a 'pure' vibration; if it had what one might call 'perfect concert pitch'.

IF THE SUN HAD PERFECT PITCH

What would things be like if the Sun were to exhibit a regular pattern of vibration, like that shown in the fourth diagram in Figure 10.6? Let us imagine that we intend to measure the rise and fall of the surface at point A, where there is an antinode, using the Doppler effect. First of all, we can obtain the frequency from the rhythm at which the motion occurs. With a second instrument we can then measure the velocity at a nearby point B, and compare the motions at the two points. If B is close to A, both points oscillate in the same sense. If the surface at A is high, the same applies at B. If B is gradually taken to lie farther away from A, it reaches a point where, in general, no motion will be measured. If B is taken to lie even farther from A, it begins to approach the next antinode. The motions of the Sun's surface at the two points are then of opposite sense. If point B is gradually taken farther and farther away, one nodal line after another is found, and from their distance apart it is possible to determine the number of nodal lines around the solar equator. In the case of the mode shown in Figure 10.6 right, there are six such points, two on each of three, separate nodal lines.

We can use a similar method to determine the number of nodal lines from North to South. In this case (Figure 10.6 right) there are two. This means that the total number of nodal lines is five and we have a fifth-degree oscillation. We now know both the frequency and the degree of the oscillation.

If the Sun does not oscillate according to the mode shown in Figure 10.6, but has an additional nodal line, its degree is six. This corresponds to a higher frequency. As with the string, the frequency increases with the number of nodes and antinodes. We can now prepare a diagram like that in Figure 10.8,

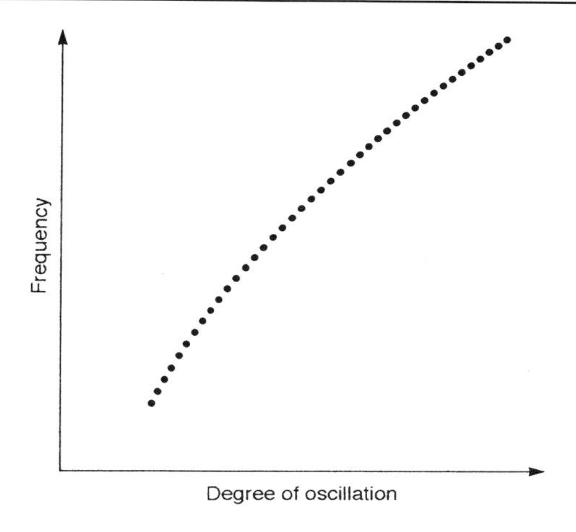

Figure 10.8. If we plot vibrational frequency of a sphere versus the degree, we find that the points form a 'string of pearls'. The higher the degree, the higher the frequency

which shows the degree of oscillation versus frequency. Even so, we are still far from understanding the oscillations of the Sun.

The degree of an oscillation is very important, because it determines the frequency even if the modes are very different. The oscillations shown in the second and third diagrams in Figure 10.6 have the same frequency, for example, because in both cases the degree is two. The simple graph in Figure 10.8 is obtained only when we consider a specific class of vibrations. As we saw in Figure 10.7, surface oscillations do not simply extend into the interior. Nodal surfaces, where the direction of the oscillation reverses, may occur at specific depths. A simple relationship between frequency and vibrational mode only applies when we consider oscillations that have the same number of interior nodal surfaces (or, to put it another way, are of the same radial order); for example, oscillations with no nodal surfaces in the interior. If we next take an oscillation mode with one interior nodal surface, we obtain a slightly different relationship between the frequency and degree of oscillation. The points corresponding to specific modes of vibration lie along a curve, similar to that shown in Figure 10.8, and resemble pearls on a string. For a given mode of surface oscillation—i.e., for a given number of nodal lines at the surface—the greater the radial order, the higher the frequency. We obtain a different 'string of pearls' for every oscillation of a specific radial order. On the graph, the strings of pearls for higher radial-order modes lie above those of lower order modes. This is shown schematically in Figure 10.9. This diagram shows that we can deduce something about the otherwise invisible solar interior from the values for frequency and degree of oscillation, which may be determined by examination of the surface. Tell me how the surface of the Sun vibrates, and I will tell you how many nodal surfaces are hidden within the interior. Let us take an example: let us assume that

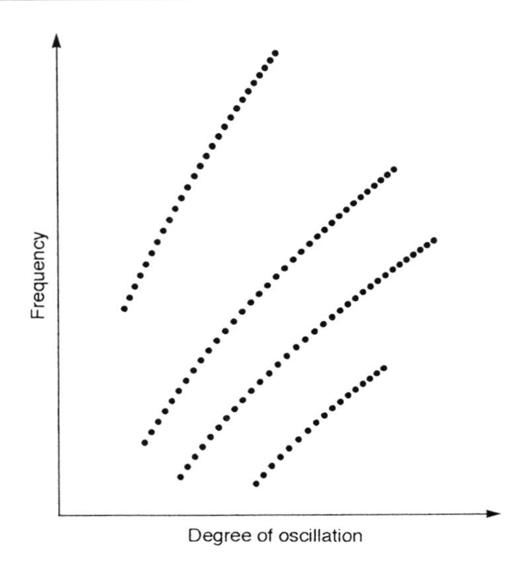

Figure 10.9. The relationship illustrated in Figure 10.8 is valid only for spheres that oscillate in the same manner in their interiors, in other words, that are of the same radial order (i.e., have the same number of invisible nodal surfaces). The 'strings of pearls' corresponding to oscillations with higher radial orders lie above those with fewer interior nodal surfaces. This means that it is possible to determine the number of hidden nodal surfaces from the degree and frequency of oscillation

we measure a frequency of 0.00276 Hz, which corresponds to a vibrational period of 377 seconds, and find that the degree is 100. From Figure 10.11—which is similar to Figure 10.9, but prepared using actual observations of the Sun—we can see that there are five nodal surfaces hidden within the interior. The diagram shown in Figure 10.11 requires some explanation, however.

MILLIONS OF MODES

We have been discussing what would happen if the Sun vibrated in a simple manner. In doing so, we have greatly simplified the situation. In fact it oscillates simultaneously in *every possible* mode, with any number of nodal lines across its surface. It oscillates in both high and low radial order modes. It is almost as if Chladni's glass plate was set into vibration, not by stroking it at one point with a violin bow, but at every point around its edges. A point that lies on a nodal line for one mode of oscillation may lie at an antinode for another mode, and thus be in continuous motion. As a result, the plate has no nodal lines: the regular oscillations have been replaced by irregular motions.

This is what happens with the Sun. There are no nodal lines on the surface and no nodal surfaces within the interior. In addition, although the modes of

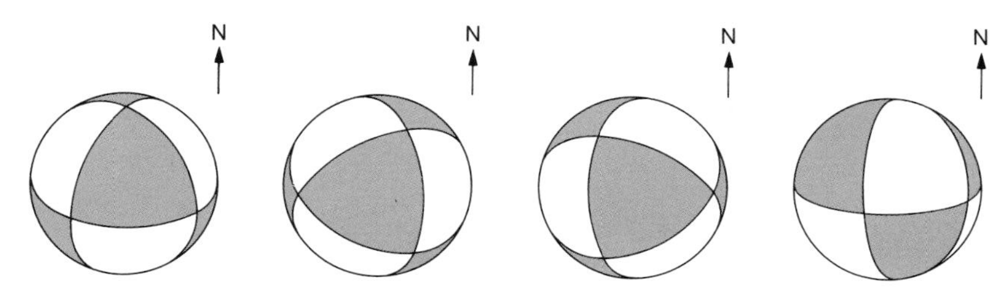

Figure 10.10. The Sun does not oscillate in a single, pure mode, such as one those shown in Figure 10.6. Its oscillations are not only of different degrees and with different numbers of nodal surfaces at depth, but also in modes with different orientations. This diagram repeats the fourth example from Figure 10.6, but shows some of the different spatial orientations that may be assumed. In the Sun, any mode may occur in any conceivable orientation. The Sun oscillates in *every one* of these modes, so that, altogether, there are millions of modes

oscillation of a sphere that we have considered had the same poles and the same equator, on the Sun any point may be the pole of a mode. So oscillations with various numbers of nodal lines and surfaces are superimposed on others, with the same number of nodes, that have different orientations, as shown in Figure 10.10. How can we possibly detect a single oscillation in this vast array of modes?

In principle it is possible to recognise individual components among a whole range of oscillations. Our ears, for example, are capable of picking out every individual instrument among a whole symphony orchestra. We are able to tell whether it was a trumpet or a violin that played a false note. We can hear if something is wrong with the flute and if this instrument's right mixture of overtones is not being produced.

The millions of oscillation modes of the solar surface may also be resolved into 'individual notes'. A mathematical procedure for doing this has been known for more than 150 years. We can, for example, take simultaneous measurements of the velocities of individual points within a rectangular area of the solar surface over a long period of time. We would then know how each point was moving at any given instant. These data may be processed, and nowadays we use a computer, of course. The program reveals how strongly individual oscillation modes with specific frequencies and number of nodal lines contribute to the overall motion.

It is not actually necessary to measure all the points within a rectangular area simultaneously. It is sufficient to sample one horizontal and one vertical strip over a considerable period of time. If one continues measuring long enough, the computer program will be able to determine the strength of the contribution from each individual oscillation. It is then possible to draw up a diagram similar to the one in Figure 10.9. Every oscillation mode that the program finds in the data has a specific frequency and number of nodal lines, which provides the position of a point on the diagram. If the Sun oscillates in numerous modes, the points must fall in a pattern similar to that shown in Figure 10.9. This was successfully shown in 1976.

THE SUN OF CAPRI

On 20 September 1974 on Capri, seeing was not particularly good for solar observations. Franz-Ludwig Deubner—who now holds the chair of astronomy at Würzburg University—was using the solar telescope at the Freiburg-based Fraunhofer Institute's* outstation on the Mediterranean island to measure velocities on the Sun by means of the Doppler effect. Little did he expect to be able to succeed in detecting very fine structure that day. It should, however, suffice for his velocity measurements, for which he intended to use a calcium line in the green region of the spectrum. The problem of the five-minute solar oscillations that had been found by Leighton and his collaborators was a matter of concern among solar physicists at that time. Were the oscillations simply caused by the rapidly rising gases in the granulation, rather like the way in which the water in a swimming pool is disturbed by the head of a diver rising to the surface? Or were they signs of disturbances in the interior? It was already suspected that all the points over the surface of the Sun did not simultaneously oscillate in the same sense, like the surface of the water in Figure 10.1b, where the pot was suspended from a spring. Theoretical considerations, in which the Californian astronomer Roger Ulrich had played an important part, suggested that the five-minute oscillations were a sign that the interior of the Sun was full of sound waves arriving from every direction, and also running in every direction, similar to the waves that arrived from every direction in Chladni's plates scattered with powdered sulphur. With such a complicated oscillation, consisting of all sorts of waves, we should not expect the solar surface to rise and fall with a single rhythm. Despite this, should it not be possible to detect oscillation modes on the solar surface, similar to those seen on Chladni's plates?

Deubner studied one area of the solar surface, and repeatedly sampled a strip, orientated north–south, with his equipment. He caused the instrument to scan towards the north for about one sixth of the solar diameter, then the same distance towards the south, to the north, to the south, etc. Taking about two minutes to do so, he repeatedly scanned the same strip of the solar surface, without a break, for nearly three hours. Then he carried out the same procedure on an East–West strip for four hours. Through the whole period, the velocity of the surface material, as indicated by the Doppler

* The Institute is now named the Kiepenheuer Institute for Solar Physics, after its founder, the solar physicist Karl Otto Kiepenheuer (see p. 53).

effect, was repeatedly measured and recorded. This wealth of data contained information about oscillations of the solar surface over the period of measurement.

Deubner let his computer analyse the data, searching for oscillation modes and their frequencies, plotting the results in a diagram similar to Figure 10.9. When we look at his diagram today—derived from data that, by comparison with nowadays, are extremely meagre—we wonder how Deubner could recognise that the frequencies and numbers of nodal lines did actually show a structure like that expected on theoretical grounds for oscillating gaseous spheres. But this is just what makes a good scientist: being able to examine one's set of data and realise, not necessarily because one is expecting something, that it contains a message, and one that will stand up to later, better measurements and scrutiny. When Deubner published his results, people at other observatories immediately began to repeat his measurements. It did not take long to confirm that the motions of the solar surface consist of innumerable regular oscillation modes. Figure 10.11 is a diagram prepared

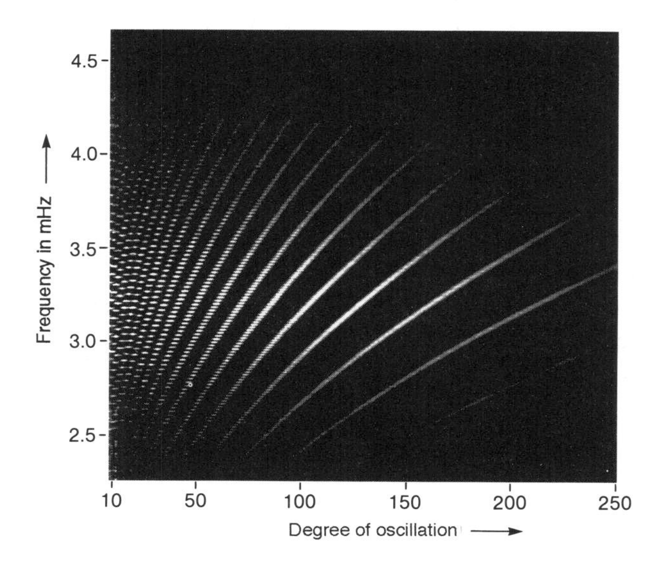

Figure 10.11. The frequency (measured in mHz, i.e., thousandths of a Hertz) and degree of observed solar oscillations. As shown schematically in Figure 10.9, the observed modes of oscillation fall on specific curves in the diagram. The frequency increases with degree along each curve. The oscillation modes represented by each curve differ in their radial order. Oscillations represented by the (indistinct) curve at lower right, for example, have three internal nodal surfaces, while those on the curve above it have four, etc. (see Figure 10.9). The diagram is derived from an uninterrupted, 50-hour, series of observations obtained by T.L. Duval, J.W. Harvey, and M.A. Pomerantz at the American base at the South Pole

from modern measurements which shows that the plot of frequency and degree does indeed appear as expected. Millions of oscillation modes have been discovered in this way, including some that produce as many as 2000 nodal lines over the Sun's surface.

INFORMATION FROM THE DEPTHS OF THE SUN

We are unable to see inside the Sun, but waves created by solar oscillations penetrate the solar interior. The oscillation modes observed at the surface therefore provide us with some information about layers hidden deep within the Sun.

In Appendix B, using a simple concept (that of a beam of sound), I have tried to explain some of the properties of the oscillations set up in the sphere of gas that is our Sun. There, for example, it is shown that the surface oscillates more than inner regions. The lower the degree of oscillation, the deeper the layers that are involved in the motion. This knowledge may be used to determine the way in which the temperature changes with depth.

The observed oscillation modes yielded a temperature profile that, by and large, agrees with previous ideas about the internal structure of the Sun. At present, however, it seems that the temperature for the centre of the Sun derived from the oscillation modes is somewhat higher than that given by solar models. That does not suit the astrophysicists. They would prefer a correction that gave lower temperatures, because it would help to solve the solar neutrino problem (see p. 157).

The oscillation modes also depend on the rotation of the Sun's interior. If it rotated like a solid body the modes would differ from those produced with a rapidly rotating core. In principle, therefore, it is possible to determine the rotational velocity of various layers inside the Sun from the oscillations. I shall describe some results shortly.

We have already seen that we can find out something about the solar interior from deep within the rock of a mountain. Solar physicists have succeeded in unlocking some of the Sun's secrets from 'observatories' that, to the general public, would appear completely inappropriate. Who, for example, would have expected that instead of the equatorial regions of the Earth, where incoming sunlight is strongest, it would be the polar regions, where the Sun hardly rises above the horizon in winter, and where even in summer it is always low in the sky, that have proved indispensable for solar researchers?

The Norwegian polar explorer Roald Amundsen's diary contains the entry for 16 December: 'It is very interesting to observe how, so to speak, day after day, the Sun crosses the sky at the same altitude. I believe that we are the first to experience this rare sight.'

SOLAR RESEARCH ON ICE

If would have been simpler if they had theodolites, but these had failed during the arduous journey. The men had to make do with sextants, but that is not easy when there is no proper horizon, unlike on the high seas. It took them several days to find the exact spot. After their journey of more than eight weeks, a few more days did not matter, however. On 15 December two men finally reached the desired spot. The third living thing was a husky. It was the middle of summer and the temperature was −19°C. The year was 1911. Amundsen and his team had reached the South Pole. A month later Captain Scott arrived. Tracks of sledges, skis, and dogs' paws in the snow, and finally Amundsen's cairn with a letter to Scott, told them that they had lost the race to the South Pole. 'Great God!' he wrote in his diary, 'this is an awful place and terrible enough for us. . . .' His disappointment may have partly contributed to the fact that he and his men died on the return journey.

Nowadays one can reach the South Pole by air, at least during the summer, and when the temperatures allow reasonably safe flights. Close to the site, only about 400 metres from the South Pole itself, scientists make use of the fact that the Sun does not set.

The time available to solar researchers is normally only that between sunrise and sunset. If, shortly before sunset, the Sun's magnetic fields eject a filament into the corona, astronomers can only find out about the fate of the ejected material if they are in touch with colleagues who are observing from a site to the west of them, where the Sun is still well above the horizon. A network of observers at various longitudes right round the Earth is able to keep the Sun under continuous observation—provided the weather is fine everywhere, of course. Recently, for example, it was possible to compile a film of changes in the solar photosphere in hydrogen-alpha light. This consisted of individual shots that had been obtained at different observatories in different time zones. The film also showed the weaknesses of this procedure. Instruments were not the same at all the sites, and differences in the weather meant that poor-quality sequences followed others that were of the highest quality, and intervening sequences were missing—because of bad weather.

The Sun may be observed without interruption from space. It is significantly cheaper and more convenient, however, to monitor the Sun from the polar regions. There the Sun does not set for half the year. It remains at the same altitude throughout the day, and circles the observer once every 24 hours.

This is why the first solar observations were made from the South Pole in 1979. An American and a Swedish scientist were able to observe events on the Sun continuously for a period of 120 hours without a break. Then French solar physicists began to do the same thing. People began to study the solar oscillations. Why go right to the South Pole? The Sun oscillates with a period of minutes, and even the few hours that the Sun is above the horizon at middle latitudes are sufficient to observe those oscillations. But if you want to find out anything from the oscillations, you *immediately* require a long observation time. We have already seen that the Sun oscillates in millions of modes of different degree and orientation. Alongside one mode of oscillation, which may have a period of 300 seconds, there is also one with a period of 300.00001 seconds. The latter is a lower degree oscillation than the former, so it penetrates somewhat deeper. From the difference between the two modes we can discover something about the layers that are reached by the second oscillation, but not by the first. It would never be possible to separate the two oscillations if observations were made over just a short period of time. If we observe the oscillations for a long time, however, they get out of step, and the computer program that filters the individual oscillation modes out of the observations is able to distinguish between them. All that is required is to observe long enough: any break in the series of measurements spoils the result.

So people go to the trouble of working at the South Pole, close to the Amundsen–Scott base, built since 1975 at a cost of six million dollars. The first observations were made in a classical manner, at almost 3000 metres above sea level, on an ice-sheet of similar thickness, and in air to which one has to become acclimatised. Like Deubner's work in 1974, the velocities of the oscillation were derived from the Doppler effect on a specific spectral line. It was soon discovered that there was an easier way of finding out something about the Sun's modes of oscillation. The rise and fall of the solar surface is linked to small variations in temperature. As we have already seen, the calcium line is a sensitive indicator of temperature. If we photograph the Sun at regular intervals through a filter that transmits the light of that line only, then it is possible to determine the mode of oscillation from the differences in brightness. The French astronomer Eric Fossat from Nice Observatory, and his American colleagues, Jack Harvey from the National Solar Observatory at Kitt Peak, and Thomas Duval Jr, from NASA Goddard Space Flight Center, therefore monitored the Sun for days on end, taking photographs of it at one-minute intervals. The computer took care of the analysis.

One of the most fascinating results of solar research at the poles concerns the rotation of the solar interior. Appendix B shows that the rotation of the inner regions of a gaseous sphere affects the mode of oscillation. Until now we knew only how the outer layers of the Sun rotate, the equator turning faster than the poles (Figure 2.8). It now appears that this rule also applies farther inside. Near the surface, where the solar energy is transported towards the outside by convection, i.e., by rising and falling columns of gas, the outer layers in the equatorial region rotate faster than those at the poles. Solar oscillations penetrate far deeper, however. They tell us that this difference becomes smaller, the deeper we go. The Sun's core, i.e., everything that is less than half a solar radius from the centre, rotates like a solid body. It takes about as long to rotate as surface regions at latitude 30°.

The science of solar oscillations is called *helioseismology*, i.e., the study of solar tremors. There is a close analogy with terrestrial seismology, the study of earthquakes. Geophysicists learn about the interior structure of the Earth from the various waves that fan out from the site of an earthquake, both those that run along the surface and also those that pass right through the Earth. In precisely the same way, solar physicists learn about the interior of the Sun from solar oscillations.

Just as we would like to know the cause of earthquakes, we also wonder how the solar oscillations arise.

A BELL OR AN ORGAN PIPE?

What is the mechanism that keeps the surface of the Sun in rhythmic oscillations? We do not know. Very probably it is a self-excited mechanism that controls the flow of energy from the interior in a rhythmic manner, sometimes allowing more and sometimes less to pass. A similar mechanism, for example, produces the oscillations of a note in an organ pipe. The air is steadily blown into the cavity of the pipe, but it emerges in pulses of alternating high and low density. The intervals between individual 'crests' and 'troughs' are the same, and this is what produces the organ's note. We know that Cepheid stars pulsate because the flow of energy from their interior is rhythmically varied by layers beneath the surface. Is this also significant in the Sun? Is the Sun an organ pipe?

Or is it a bell? The sound of a bell is produced by the clapper, which repeatedly strikes the body of the bell. Each time it emits a note, and before that note has died away, it is set into vibration again by a further stroke. In the Sun the irregular motions of the granulation cells could play the part of innumerable clappers. The rising and falling columns of gas may give a continuous series of blows to the body of the Sun, keeping it in oscillation. At present it seems as if the betting is in favour of the bell model.

11

THE RADIO SUN

> The great storms of radio emission from the Sun in February 1942 marked the beginning of the modern development of radio astronomy. When British Army radar stations experienced severe jamming during late February, an investigation made by J.S. Hey (the author) led him to conclude that radio waves of amazing intensity were being radiated by the Sun, apparently due to the presence of a very large and active sunspot on the solar disk.
>
> J.S. Hey, *The Radio Universe*

By 7 o'clock Central European Time the flotilla had reached the longitude of Cherbourg. Vice-Admiral Otto Ciliax was pleased. They would soon make up for the two hours delay. But the most difficult part of the journey still lay ahead of the three battleships. Only three hours after sailing from Brest had the crews of the *Scharnhorst*, the *Gneisenau*, and the *Prinz Eugen* been told the aim of this operation, ordered by Hitler. The three battleships were on their way through the English Channel, heading for Wilhelmshaven, to be deployed in the North Sea to protect the transport of ore from Norway to Germany. As yet the British radar system had not detected them. In fact it was not until 13:18 that the flotilla was noticed. By that time it had passed through the Straits of Dover. The attacks that followed were unable to prevent the operation, which was known by the code-name 'Cerberus', from reaching a successful conclusion. The ships reached their German ports of destination as planned. On 12 February 1942, the British radar was a failure.

The Germans later extolled the careful preparations that they had made, by repeatedly transmitting jamming radiation previously, so that no British suspicions were aroused when there was strong radar interference during the hours that mattered on 12 February. Did the operation succeed because the Germans jammed the British radar? Shortly after the three ships had broken though the blockade in the Channel, Winston Churchill blamed the failure on 'atmospheric disturbances'. A few weeks later, the British radar system was again jammed. Were the Germans about to launch an invasion? There was a major state of alert, but no attack followed. In the meantime, a young physicist, J. Stanley Hey, had begun an investigation. It did not take him long to find out that the jamming was not of German origin, but came from the Sun.

Since then, we have come to realise that the Sun does not emit light and heat alone, nor just the masses of gas that escape from coronal holes and stream past the Earth. The Sun also emits a whole series of radio signals. The person who discovered the Sun's radio emission, Stanley Hey, who had previously been concerned with the physics of crystals, never lost interest in this new field and eventually became a respected radio astronomer.

THE RADIO DISH ON THE SCHOOL ROOF

In 1932, a radio engineer in the USA, Carl Jansky, accidentally discovered that radio waves were arriving from space. It was only in 1942, with the war, when the aether was systematically scanned for radio waves, that this was remembered. Nowadays, pupils studying physics are able to listen in to the Sun's radio emission.

Anyone who happens to see the equipment on the roof of the Saint Michael Gymnasium in Bad Münstereifel, thinks that the German post office has erected it to transmit telephone conversations via satellite. A radio dish, 1.75 metres in diameter, stares at the sky. Anyone who looks more closely, however, will see that the dish is movable, and follows the Sun in its daily motion across the sky. It is being used to measure the radio emission at a wavelength of 11.1 cm.

It is not by chance that this school in Bad Münstereifel has a radio telescope on the roof. Close by, there is the giant radio telescope belonging to the Max-Planck Institute for Radio Astronomy, whose dish has a diameter of 100 metres. Where 'big research' is carried out, there is always a fallout of 'mini-research'. Despite help from their teachers, the amateurs had to work hard at their small aerial on the school roof. Anyone who manages to borrow a parabolic aerial and a chart recorder is still far from having a radio telescope. The pupils, guided by physics teacher, Walter Stein, had to fix the aerial to an axis, so that it could rotate. An old electrical motor was repaired, and counterweights—taken from a piece of body-building equipment—were provided so that the parabolic dish would be light enough on its bearings for it to track the Sun. The dish itself was hauled up to the roof with ropes and baulks of timber. It was then connected to a driving mechanism, built by the pupils, and also to the electronic equipment to detect and measure the strength of the radiation.

The radio telescope was ready by the beginning of July 1987. It was able to track the Sun, and the chart recorder registered the strength of the solar radiation. The school holidays began, but the chart recorder did not show any deflections. This was not surprising, because solar activity was at minimum in 1987. Day after day the pupils stared at the recording, but nothing happened.

Finally, on 24 July, at 12:00 Universal Time, they were rewarded: the pen was deflected. In later years there were more solar eruptions, which were recorded.

One of the events recorded by the pupils at Bad Münstereifel is shown in Figure 11.1. Mr Stein did not simply let his pupils record radio bursts, however. He also encouraged them to think about where the Sun's radio emission originates.

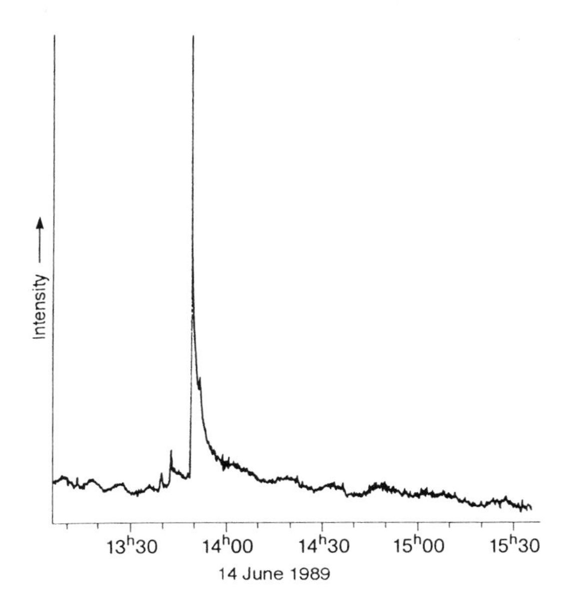

Figure 11.1. The flare of 14 June 1989 as recorded by physics-group students at the St Michael Gymnasium in Bad Münstereifel

AN AGITATED PLASMA

Where does the Sun's radio emission come from? It originates in just the same way as in a radio station. A radio station's aerial is a metallic electrical conductor. The electrons that fill the spaces between the metal ions are able to move freely. The transmitter forces them to move rhythmically backwards and forwards along the cable to the aerial. The moving electrons produce a current, which, in response to their motion, repeatedly reverses its direction. Like any current, the alternating current in the aerial is accompanied by a magnetic field. Again, the changes in the direction of the current cause the magnetic field to reverse repeatedly. In a typical radio station's aerial this happens more than a million times a second. When a magnetic field changes very rapidly with time, electromagnetic waves of the sort that we discussed in Chapter 3 are produced. Radio waves are just the same as light waves, except that their wavelengths are much longer. Instead of amounting to ten-thousandths of a millimetre, they range from millimetres to hundreds of metres in length. The radio waves produced by the aerial propagate through space at the speed of light.

In a plasma, conditions are similar to those found in an aerial. Once again, the electrons are free to move relative to the ions. Imagine that a cube of plasma is moving through space. The electrons and ions are moving at the same speed. Now let us assume that all the ions are suddenly braked and brought to a halt. Initially, the electrons continue to move unhindered in their original direction, because at first they are not retarded in any way. In doing so, however, they move away from the ions. As a result, shortly after the braking began, there are more electrons in the front half of the cube than previously. The front therefore acquires a negative electrical charge. Correspondingly there is a dearth of electrons in the rear portion, where the positive ions predominate. The electrical charges exert forces on the electrons. The negative forward side repels them, and the positive rear side attracts them (Figure 11.2). The result of all this is that the forward motion of the electrons is reduced, and they are pulled backwards. Once they are moving backwards, they do not move just far enough to compensate for the charge on the positive ions in the rear. They overshoot their goal, so the rear becomes negatively charged, and the front positively charged. The electrical forces created by this new charge separation pull and push the electrons forwards again. So the

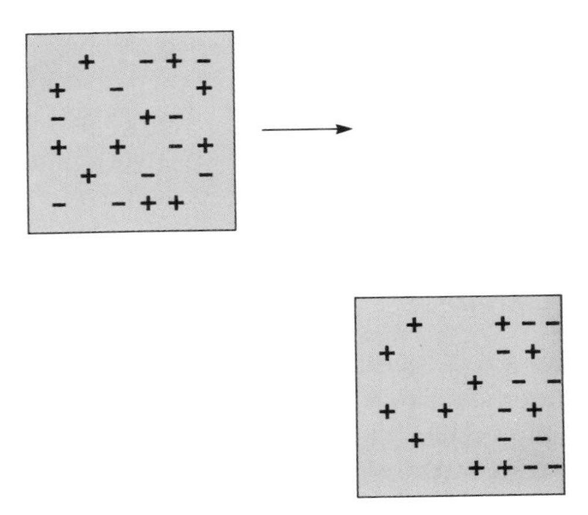

Figure 11.2. Top: In a plasma, shown here moving towards the right, the positively charged atomic nuclei (+) and electrons (−) are moving with the same velocity. The opposing charges maintain their mutual positions and the plasma is electrically neutral. Bottom: If the atomic nuclei are suddenly braked, initially the electrons continue to move unhindered. The right-hand portion of the plasma, therefore, contains a concentration of negative charges, while the left-hand portion contains mainly positive charges. The opposing charges attract one another, and the electrons are pulled back towards the left. This is how plasma oscillations are initiated

electrons continuously oscillate between the front and the rear, producing an alternating current in the plasma, just as happens in an aerial. The current itself produces an alternating magnetic field. The oscillations of millions and millions of electrons relative to the more massive, and thus almost motionless ions, are known as *plasma oscillations*.

Normally, the strongly attractive forces between the negative electrons and the positive ions ensure that a plasma is neutral. If, somewhere, there is an excess of positive charges, they attract electrons from the surroundings, whose negative charges neutralise the excess positive charge. If a plasma is left to its own devices, it becomes electrically neutral. If the electrons and ions move relative to one another, perhaps in response to external influences—as with the sudden braking of the plasma in our example—then the charge equilibrium may be destroyed. The powerful electrical forces then attempt to re-establish the neutral state, and the electrons begin to oscillate relative to the ions.

Because this involves the motion of electrical charges, currents and magnetic fields are induced. The frequency of the backwards and forwards oscillation of the electrons is known as the *plasma frequency*. The greater the electron density, the higher the frequency. In the solar corona the plasma frequency is around ten million oscillations per second. This produces radio waves with wavelengths of 30 metres. Because of the higher electron density close to the surface of the Sun, there the plasma frequency is about 100 million oscillations per second. The corresponding radio waves have wavelengths of a few millimetres or less.

Electrons emit radio waves not only when they undergo regular oscillations but also when they move erratically, such as when they encounter some hindrance to their motion. This may happen, for example, when an electron encounters an ion, i.e., an atom that is lacking one or more electrons. The attraction exerted by the positive ion on the negative electron deflects the latter from its straight-line trajectory. Depending on how close the two particles come, and how fast they are moving relative to one another, the electron is braked to a greater or lesser degree. Any change in its motion causes it to emit a quantum of radiation (a photon). It soon encounters another ion or an electron. Again it is deflected. As a result it continuously emits radio waves. In every gramme of hot solar material, millions and millions of photons are emitted every second. Because the gas in the solar atmosphere is poorly transparent to radio waves, however, not all the radiation that is produced reaches us.

RADIO SILENCE AT LONG WAVELENGTHS

A radio station's aerial is not surrounded by a plasma but by electrically non-reactive air, so the radio waves that it emits move through space at the velocity of light. In other words, they cover 300 000 kilometres every second.

In a plasma, things are different. This is because it contains free electrons, which may be excited into oscillation by any incident radio wave. If there were no free electrons in the gas surrounding the transmitter (as with a terrestrial radio station), radio waves of any wavelength, however short, would be able to escape.

The plasma frequency that we mentioned earlier plays a decisive role here. No radio waves with frequencies lower than the plasma frequency are able to propagate through a plasma. If we use the relationship between wavelength and frequency (which is explained in Appendix C), we can say that in a plasma, only radio waves whose wavelength is below a certain critical limit are able to propagate. The higher the electron density, the higher the plasma frequency, and thus the lower the limiting wavelength. Close to the surface of the Sun, only emissions with wavelengths of less than one millimetre are able to propagate. They are able to reach us because the overlying layers have a lower electron density, and thus a lower plasma frequency and a higher critical wavelength. Anything that is able to escape from the lower layers has no problems in penetrating the outer layers of the solar atmosphere, and eventually reaching us. We receive radio waves with wavelengths of 10 centimetres from just the outer atmosphere. Waves in the metre region reach us only from the corona.

Radio images of the Sun provide direct evidence for the remarkable properties governing propagation of radio waves in the solar atmosphere.

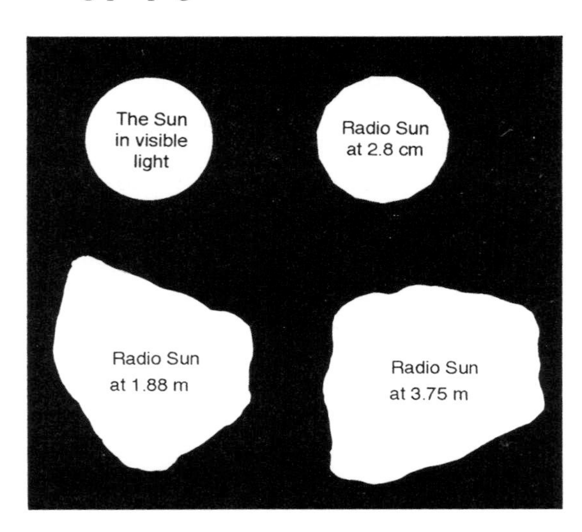

Figure 11.3. Observed with a radio telescope, the Sun appears different sizes, because the longer the wavelength (and thus, as shown in Appendix C, the lower the frequency), the farther out in the corona are the regions from which radio waves may reach us. The diagram shows (very schematically) the outline of the Sun at various wavelengths. Because of the variability of the corona with time, the radio image of the Sun also varies considerably depending on when it is observed

Figure 11.3 shows how the Sun appears in the visible region. A radio telescope using wavelengths of less than one millimetre would show a similar-sized image. The diagram also indicates schematically the size of the radio image at various longer wavelengths. The telescope 'sees' a larger Sun, the longer the observed wavelength, because it is able to penetrate only into the corona. At a wavelength of five metres, only radio waves from the outermost layers of the corona are able to reach a radio telescope.

WAVELENGTH ALLOCATION ON THE SUN

The density of the solar atmosphere, like the Earth's, decreases with increasing height. Correspondingly, the number of free electrons per cubic centimetre is far larger at the bottom of the solar atmosphere than it is at the top. A 14-figure number is required to express the number of electrons per cubic centimetre close to the surface, but just seven figures are needed for the electron density in the corona. We have just seen that for a given electron density, only waves below a certain critical length are able to propagate. Let us imagine that we had an ordinary terrestrial radio transmitter on the surface of the Sun. (We can ignore the minor inconvenience that at a temperature of 5000°C, any terrestrial transmitter would be vaporised immediately.) At a wavelength in the medium-waveband—say the 194.4 m of the German radio station that I listen to—the waves would be unable to propagate though the plasma near the solar surface. Even if we used a VHF transmitter, no one would be able to pick up the signal, because the wavelength would still be a few metres—still far too long. Only if we were able to transmit at a wavelength of about one centimetre would there be any chance that someone might hear us. If, instead of being down on the photosphere, however, our transmitter were at an altitude of 4000 kilometres, we would succeed at a wavelength of just 50 cm, because there the density of free electrons is so much lower. To transmit signals in the metre band we would have to be much higher. If we wanted to capture signals in the VHF region that is used for terrestrial radio services (say at a wavelength of 2.7 m or 111.1 MHz), the transmitter would have to hover at least 70 000 kilometres above the Sun's surface. Such a distance amounts to about one twentieth of the Sun's diameter.

RADIO BURSTS FROM THE SUN

Radio observations of the Sun from Earth clearly show the effects of the properties just described on the propagation of radio waves.

Radio bursts from the Sun originate in flares and other eruptions. When material is flung into space, it excites oscillations in the surrounding plasma. Electrons and ions oscillate at the plasma frequency, just as in our thought experiment with a cube of plasma. That creates alternating magnetic fields. Such waves propagate through the plasma and reach us, because their frequency is precisely the local plasma frequency.

This is the cause of a remarkable effect: When there is an outburst, only short wavelengths reach us initially, but as time passes, we receive radiation at longer and longer wavelengths. The reasons for this are, of course, the properties governing radio-wave propagation that we have just discussed. Assume for a moment that the ejected clump of plasma is at a height of 4000 km. Only waves below 66 cm in length can reach us. If the clump of plasma continues to rise and later causes the plasma at a height of 70 000 km to oscillate, waves with a length of less than 2.7 m are able to reach us. This enables us to determine the velocity of the clump of plasma that was ejected. If waves in the metre band—to be precise, at 2.7 m—reach us 73 seconds after 66-mm waves arrive, then the clump of plasma covered the difference in height of 66 000 km in 73 seconds, i.e., at a velocity of 900 km/s.

That is a very leisurely velocity. Bursts of this type are known as Type II, in which the plasma is ejected at relatively low velocities. In Type III bursts, the ejected swarms of electrons reach velocities of nearly 100 000 km/s. Figure 11.4 shows schematically how the wavelengths that are detected change with time.

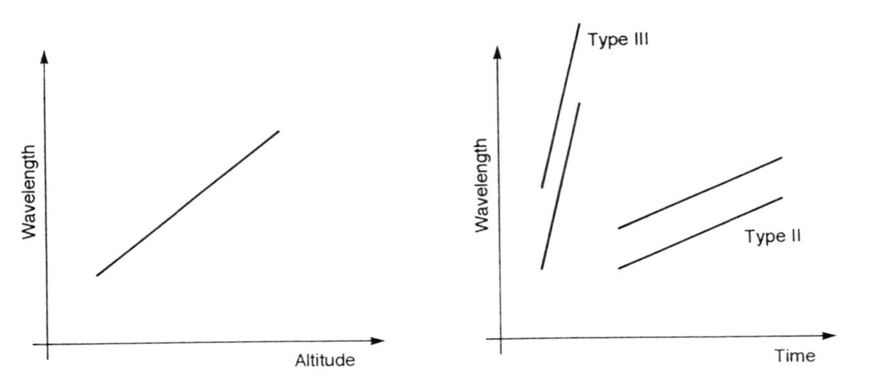

Figure 11.4. Left: Critical velocity versus altitude above the solar photosphere. Waves that have wavelengths longer than the critical value are unable to escape. The greater the altitude, the longer the radio waves that are able to escape into space. Right: When material collides with the plasma in the solar corona, radio waves are produced. As the material exciting the coronal plasma rises, longer and longer waves are able to escape and reach us. In a Type III burst, the disturbance rises extremely fast—the wavelengths of the radiation that we detect rapidly become longer. In Type II bursts, the rate of ascent is considerably slower, and thus the increase in wavelength is much less. Because the only frequencies that are produced in a burst are those of the plasma oscillation and its first overtone, radiation of two different wavelengths is detected at any one time

THE UNIVERSE AS REFLECTED BY THE SUN

The propagation properties of radio waves in a plasma cause any radio wave incident on the Sun to be reflected by the solar corona. Look at Figure 11.5. We have seen that the electron density of the corona increases downwards. The velocity with which a wave-crest propagates through a plasma, however, increases the more free electrons there are in a given volume of space.* We can apply similar considerations to those used in Appendix B for sound waves propagating within the solar interior. Let us examine a wave that is incident on the corona from outside in more detail. The 'crests' of the magnetic field—which we may interpret as successive 'wavefronts'—are marked in the diagram by short transverse lines. Because the velocity of the wave crests is greater in underlying layers than it is in higher ones, the portion of the wavefront nearest to the Sun moves faster. As a result, the wave changes direction. Any radio wave arriving from space is deflected. It is reflected in just the same way as a beam of light that falls on a reflecting sphere. If the corona were perfectly spherical, we would be able to see, reflected in it, the image of every cosmic radio source. The whole sky would be reproduced in the reflected image, just as a Christmas-tree ornament reflects the whole of one's living room.

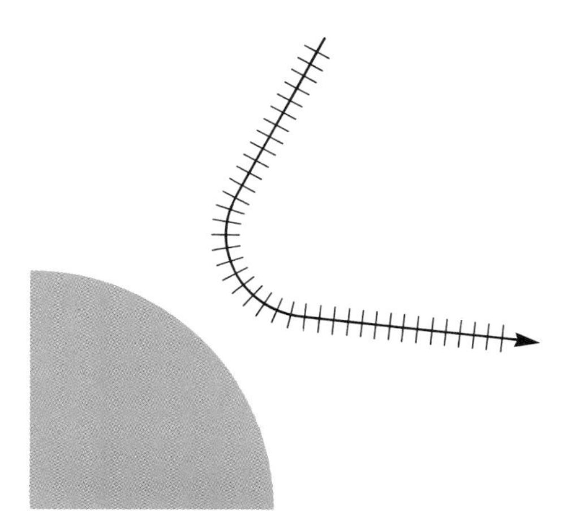

Figure 11.5. Radio waves incident on the solar corona from outside curve back into space, because the portion of each wavefront (transverse lines) that is closer to the Sun moves faster than the portion that is farther away. Similar considerations also apply to sound waves that are propagated inside the Sun (see Figure B.2)

* For the sake of simplicity, we will consider here just the velocity with which one wave crest follows another, known as the *phase velocity*. In a plasma this may be greater than that of light in empty space. This does not contradict Einstein's dictum that nothing may exceed the velocity of light. We can appreciate that they may contain wave crests that move faster if we watch an expanding wave on water. Within the wave, there are faster wave crests and troughs, which originate at the inner edge and disappear at the outer.

The corona is not homogeneous, however. It is crossed by magnetic fields that are anchored in the Sun. Coronal holes with a lower electron density alternate with regions of higher electron density. As a result we are unable to see a reflected image of the radio sky.

HERR MEYER AND THE RADIO ASTRONOMER

After the lecture, the audience was able to pose questions. At first none of the public had anything to say. Then someone plucked up courage, and broke the ice. A professor from the University of Bonn had been speaking about his field of research, radio astronomy, and particularly about radiation from space. Herr Meyer had been able to follow the lecture reasonably easily, except at the very end, when the speaker touched upon radio waves from the Sun. Here, Herr Meyer became confused. The radio astronomer had described so many types of radio burst that Herr Meyer could not tell one from the other. He did not want to make a fool of himself by asking, however. Perhaps he was the only one who had not understood. Later, as the audience was leaving, he made his way to the front, and asked the scientist about the different types of burst.

'Have I understood correctly that a burst at metre wavelengths begins with a Type I burst, followed by a Type II, and then finally a Type III?'

'No, the burst begins as Type III, and Type II follows.' With that, Herr Meyer though he had finally understood the numbering scheme.

'So if Type II follows Type III, the third is then Type I?'

'No', the professor said, 'Type IV comes next.'

'So the order is always III, II, and IV?' asked Herr Meyer.

'Not always. In the centimetre region, IV succeeds III, because Type II bursts can hardly be detected at short wavelengths.'

Herr Meyer was now thoroughly confused.

'And when does Type I occur?'

'Type I has nothing to do with the other types. Type I is encountered in the middle of a radio storm.'

Herr Meyer gave up.

On the way home he was thoroughly annoyed. Even when he got back home and pulled a book on radio astronomy from the shelf, he found he was unable to bring any sense of order to the different types of radio bursts observed on the Sun. He suspected that the numbers must have originated when bursts were first discovered, before anyone had an overall picture of what was happening. But that did not get him any further.

Somewhere in his subconscious he must have retained the significant points, however, because of the dream that he had the following night.

'The storm has now lasted for two days', the professor was saying, standing alongside him. Herr Meyer looked out of the window and saw a large dish aerial. Obviously he was at some radio observatory. It was a warm summer's day and the Sun was blazing down. There was not the slightest breath of wind to stir the leaves on the trees.

'I mean the radio storm on the Sun. We are measuring its radio waves at metre wavelengths', explained the professor, when he saw Herr Meyer's puzzled face.

'It's invisible, of course. You can't see it, but you can hear it.' He turned a knob on one of the pieces of equipment in front of him, and from somewhere a rushing noise filled the loudspeakers. Instinctively, Herr Meyer thought of the sound of waves on the sea or of a waterfall. Suddenly, for a few seconds, the noise became louder. After about a minute, the louder sound was repeated (Figure 11.6).

'Those are Type I bursts, which accompany a radio storm. Since yesterday the Sun's radio emission has been about one thousand times stronger than it was last week. I expected a storm, because for the past few days a large sunspot group has been crawling across the middle of the disk. See for yourself.'

He led Herr Meyer up some steps and into a small dome, in the centre of which there was a telescope. A slit in the dome allowed sunlight to fall on the telescope.

'Look through the eyepiece', he said. When Herr Meyer did so, he could see the disk of the Sun, held motionless in the field of view as the telescope automatically tracked the Sun. The Sun's brilliant light was heavily filtered to prevent it from damaging the eye. Herr Meyer saw a group of sunspots right in the centre of the disk.

'Regardless of what is producing the radio waves', the professor explained, 'they appear to propagate upwards perpendicular to the surface. When a

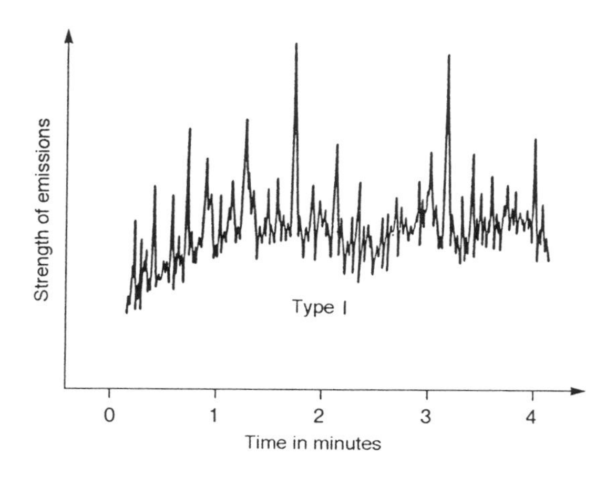

Figure 11.6. A radio storm on the Sun, which may last for hours or days, shows abrupt bursts of Type I at metre wavelengths

spot appears in the centre of the disk as seen from Earth, we intercept the radiation. In the next few days the radio storm will decline, because the Sun's rotation will carry the spots farther round, and we shall no longer be in the line of fire. Probably what we have there is intense particle radiation, which is exciting plasma oscillations in the corona. The radio emission is at metre wavelengths, which can reach us only from the corona. Even spots that are not at the centre of the disk can affect radio emission. This was shown when the Canadian radio astronomer Arthur E. Covington observed a partial eclipse of the Sun in November 1946. He followed the Moon's motion across the Sun and noted that as spot after spot disappeared, so did the radio emission coming from each spot. He realised that the spots made a significant contribution to the Sun's overall radio emission, especially in the short-wave region. Near sunspots, the Sun's emission was stronger than in quiet regions of the photosphere. This is particularly noticeable during the practically spot-free period around sunspot minimum. Then the emission is just that of the quiet Sun.'

'Where do the quiet Sun's radio waves come from?' asked Herr Meyer. In the meantime they had returned to the control room, where the rushing noise of the Sun from the loudspeakers drowned out nearly everything else.

'They simply originate as radiation emitted by electrons that zigzag through the gas and collide with other particles. In the metre waveband its emission comes from electrons in the corona, and at shorter wavelengths from electrons in deeper layers, whose longer-wavelength radiation cannot escape. Because the corona is not homogeneous—there are regions of high electron density, and others, the so-called coronal holes, where it is much lower—emission from the quiet Sun is also patchy. But if there are sunspots, the corona appears to become denser above them. There are more free electrons per cubic centimetre in those regions (known as 'condensations') than there are elsewhere. Where there are a lot of electrons, a lot of radio emission is produced. So, apart from the background radiation from the quiet Sun, there is what is termed the 'slowly varying component', which originates from condensations in the corona above sunspots. Because the spots are carried round by the solar rotation, taking, on average, about 27 days to complete one rotation, the radiation also varies with the same period. We do not receive any radiation from condensations on the far side of the Sun.'

By this time Herr Meyer had become accustomed to the Type I bursts of the on-going storm. For a few seconds every one or two minutes they overrode the rushing noise of the storm. Suddenly, he pricked up his ears. It sounded as if a breaker was racing towards the shore. The noise became louder and louder, reached a peak, and ebbed away again. A few minutes later, the ordinary rushing noise had returned. Before long, however, the noise began to strengthen again. This time it drowned out the storm for a quarter of an hour, during which time its strength fluctuated very rapidly. Then it disappeared, leaving just the ordinary storm noise. Shortly afterwards the noise-level rose again.

'The first was a Type III burst, lasting just a few minutes. Our equipment is not capable of recording the preceding microwaves, being only suitable for wavelengths around one metre. Then there was the long Type II burst, and after it had decayed, the Type IV, which you can still hear. It will last a bit longer (Figure 11.7). A pity we were not looking through the telescope, because we might have been able to see a flare. Whenever there is a flare there is a Type III burst. The subsequent Type II and Type IV bursts are, so to speak, the thunder after the lightning. The energy that is released in a flare, and partly converted into radio waves, probably comes from magnetic fields that annihilate one another. In a Type III burst, which begins simultaneously with the flare, electrons are ejected at velocities close to that of light. They cause oscillations in the plasma, creating the radio waves that we observe.

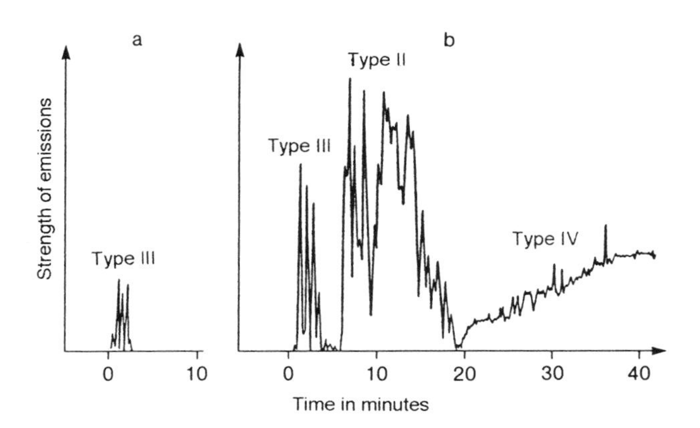

Figure 11.7. The radio emissions from flares on the Sun. Left: A small flare is accompanied by a Type III burst. Right: A strong flare, on the other hand, initially causes a Type III burst, then a Type II, and finally a Type IV. Figure 11.4 shows how the Type III and Type II bursts may be distinguished on the basis of the speed at which they change

Frequently that is all the radio emission that there is from a flare. With strong flares, however, it is by no means all. One of my colleagues once said that the Type III burst and the microwaves that accompany a weak flare can be compared with a volcanic eruption, where only sparks and ash are ejected. With a stronger flare, however, Type II and Type IV bursts occur. They are the lava. In the subsequent Type II event, the ejected material moves more slowly, although still at supersonic velocities. It moves through the corona at perhaps just 1000 km/s. If it encounters plasma, oscillations are induced in the latter. The final, Type IV burst may be observed at all wavelengths.

You should think of the process as something like the following:* A flare erupts on the surface of the Sun. Vast amounts of energy, which were previously stored in magnetic fields, are released. Whilst everything is happening at the base of the corona, we observe only short-wave radiation. Simultaneously, however, streams of electrons are ejected upwards. Their velocities are around 100 000 km/s. As they rise, they excite higher and higher regions of the corona, and longer and longer wavelengths reach us. This is a Type III event. Matter ejected by the event (a giant clump of plasma) then follows with a velocity of only some 1000 km/s. Its forward edge collides with the stationary material in the corona, and excites oscillations in the plasma. That is the Type II event (see Figure 11.7). As it rises, the ball of plasma hauls part of the Sun's magnetic field inside it up into the corona. Electrons within it and the oscillations of the shock-excited plasma produce yet more radio waves. The higher the clump of material rises, the longer the wavelength of the radiation that reaches us. This portion of the outburst is designated a Type IV event. At centimetre wavelengths it begins earlier than at metre wavelengths. Once again, that is because when it is still at low levels, long-wave radiation cannot escape.'

When Herr Meyer woke up the next morning, he felt that his nocturnal tutorial had actually enabled him to understand a bit more about the Sun's radio emission.

BOUNCING RADIO WAVES OFF THE SUN

For millions of years the Sun has been sending radio signals to the Earth. In September 1958 we answered. In fact, we received radio echoes from the Sun. As we saw on p. 195, the solar corona reflects radio waves. So we can expect radio signals reaching the Sun from Earth to be reflected back into space.

The radar aerial at Stanford University in California consisted of four individual elements, which were spread over a rectangular area of about five

Figure 11.8. How the radio emission from a flare may be explained. Initially, only short-wave, microwave radiation can reach us from the base of the corona (a). The energy released by the flare shoots high-velocity electrons up into the corona, where they excite oscillation in the surrounding plasma. A Type III burst is observed, with its characteristic increase in wavelength, which indicates that the electrons are moving at about one third of the speed of light (b). The energy that is released also propels a clump of plasma up into the corona. At its leading edge it excites oscillations in the coronal plasma, which then emits the radiation observed in Type II bursts. Its wavelength increases more slowly than in the case of Type III bursts. The magnetic-field lines frozen into the plasma are forced to follow the clump of gas as it rises. They excite oscillations in the plasma at the base of the corona, which then emits short-wave radiation (c)

* The three successive stages described by Herr Meyer's professor are shown in Figure 11.8. Figure 11.7 shows the events as they appear at metre wavelengths, as observed from Earth.

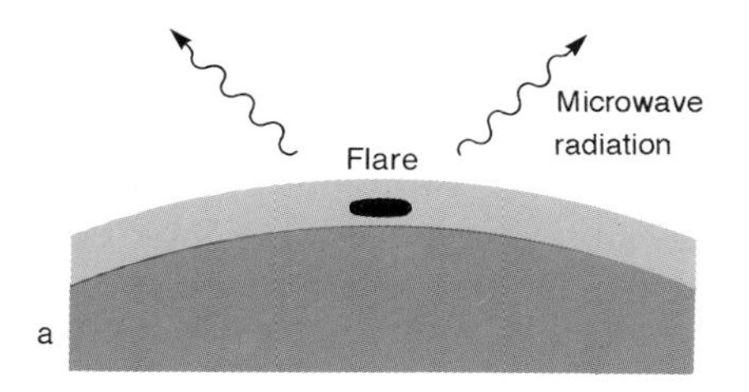
Microwave
radiation
Flare
a

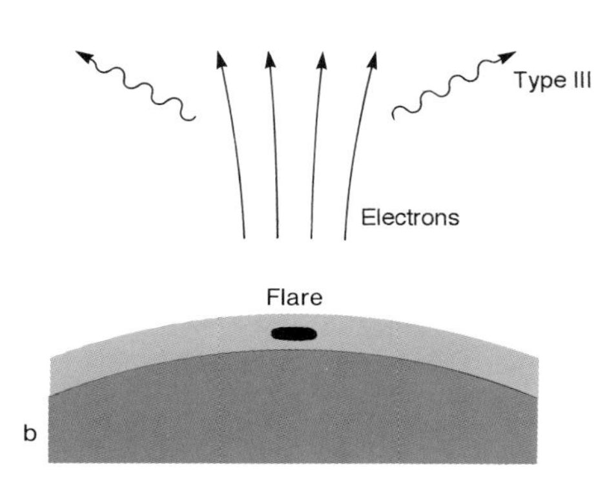
Type III
Electrons
Flare
b

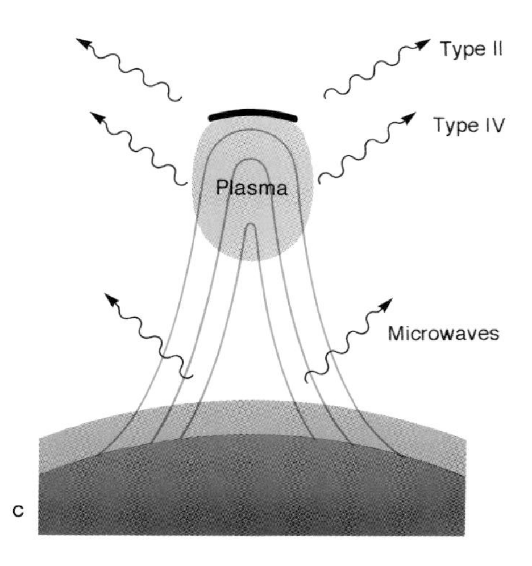
Type II
Type IV
Plasma
Microwaves
c

hectares. Because the installation could not be moved, the Sun was rarely within the field of view. For just a few days in the year, in April and September, the Sun lay within the radar beam, which pointed towards the East, for about 30 minutes. The first attempt to capitalize on this fact was in September 1958. Radar signals at a wavelength of 11.7 m were beamed at the Sun. The message transmitted was extremely simple. For 30 seconds, a steady signal was transmitted. Thirty seconds of silence followed, then another 30 seconds of signal, etc., for a total of 15 minutes. Then the transmitter was switched off, and a receiver connected to the array.

The time of 15 minutes was not chosen at random. A signal, moving like the radar signal at the speed of light, takes about eight minutes to reach the Sun from Earth. The same amount of time is required for the return journey. So the first radar echo from the Sun could be expected about one minute after the switch-over—if everything went well. In principle, it should have been possible to hear the whole transmission over the previous quarter of an hour as an echo: 30 seconds of signal, 30 seconds of silence, signal, silence, etc.

It did not turn out to be as simple as that. The Sun emits radio waves itself, including ones at the same frequency as those being produced by the radar installation. This unwanted radiation meant that it was difficult to detect the echo. The bulk of the signal consisted of radio waves created in the outermost layers of the corona. It was difficult to detect the faint echo of the human signal against that background. It is as if, among all the hubbub of a Munich Oktoberfest, one was trying to hear what an acquaintance, sitting several tables away, was calling out.

Using modern statistical methods, however, it is not only possible to clearly make out the echo, but also to determine how the reflection altered. If the reflecting material is in motion, the Doppler effect alters the original frequency. If the material reflecting the signal is approaching the radar aerial, then the echo has a shorter wavelength than the original signal. Conversely, if it is moving away, the echo has a longer wavelength. Echoes from the Sun, however, come from material rotating with the Sun. This rotation means that the radar signal illuminates both material that is moving away from us, and some that is moving towards us. Part of the echo then consists of longer, and part of shorter wavelengths than the original signal. The echo therefore also contains information about the rotation of the solar corona.

In addition it also proved possible to determine something about motions in the corona itself from the radar echoes. We have already seen that material in the corona is flowing outwards from the Sun along the magnetic lines of force. Throughout the corona, there is thus a uniform wind direction from bottom to top. Matter is flowing outwards. The radar echoes are also affected by this motion. On average, they shift towards shorter wavelengths, which is itself a sign that it is material flowing towards us that is reflecting the signal. This method has enabled us to measure the velocity of the solar wind in the corona, and this has been found to be at least 20 km/s.

12

TO THE SUN

> It was a simple job, really: flying a specially made energy-collecting spacecraft to the sun. The dipper flew through the corona and chromosphere to the relatively cool surface of the photosphere. . . . Collecting cells in the dipper's ship absorbed all varieties of energy—heat, light, gravity. . . . Banks of accumulators, sheathed in delicate magnetic 'bottles' held the energy until it could be transferred to purchasers waiting in Mercury orbit.
>
> Paul B. Thompson, *Stardipper* (1987)

In contrast to this quotation from a science-fiction story, our attempts to approach the Sun have been both tentative and modest.

THE BIRTH OF SPACE RESEARCH

Every ray of light that reaches the Earth's surface from space has had to pass through our atmosphere. From the whole range of electromagnetic radiation, only visible light and radio waves are able to penetrate this sea of air and reach the ground. Little of the ultraviolet or infrared light, and none at all of the X-rays or the even more energetic gamma-rays succeed in reaching us. Until the end of the Second World War therefore, nothing was known about the radiation emitted by the Sun that is blocked by the Earth's atmosphere.

The beginning of space research using measurements made outside the Earth's atmosphere may be dated to the very day. On 10 October 1946, a V-2 rocket (a captured German weapon), was launched from White Sands in New Mexico. It carried measuring instruments to an altitude of 90 kilometres. During the short period when the rocket was above the atmosphere, photographs were taken of the solar spectrum in the extremely short-wave ultraviolet region. This radiation is absorbed by the uppermost layers—at least it is as long as molecules of propellant from our spray cans have not reached that height. Our atmosphere has not quite come to an end, 90 kilometres above sea level, but only about one millionth of the Earth's atmosphere remains above that height. That original flight was therefore able to measure radiation that never reaches the Earth's surface.

In subsequent years the measuring instruments were improved, and the V-2 rockets, which had been built for the aerial bombardment of London, were replaced by new launchers. Nevertheless, only a short time was available at each launch, when the rocket was near the peak of its trajectory.

A BALLOON-BORNE OBSERVATORY

Solar researchers would like to site their telescopes on islands and on high mountains. They would like to observe from white-painted towers that extend above the turbulent layers near the ground. Mostly, however, they would like to be free of the problems caused by the Earth's atmosphere. This was realized by Martin Schwarzschild, a professor at Princeton University, who, in 1953, working from Mount Wilson Observatory in California, began to study the granulation, the continuously varying, fine structure caused by motions in the Sun's outermost layers.

People were just beginning to use the rockets to study, in space, the types of radiation that could not reach the Earth's surface. They were breaking new ground. Because visible light reached the surface anyway, no one thought of using expensive rockets to study it. As we have seen in Chapter 4, however, even visible light is adversely affected by passing through the Earth's atmosphere. The plan that was devised in Princeton envisaged carrying instruments into the upper layers of the atmosphere on board a balloon, and observing the solar granulation from that altitude.

Observations of the granulation were affected by the constant motions in the Earth's atmosphere, which prevented observers from detecting fine structure on the Sun. Better telescopes and better photographic techniques had not been able to improve matters. In 1957, the best photograph of the granulation remained one that had been taken in the previous century. On 1 April 1894, Janssen—we encountered this pioneer French solar observer in Chapters 4 and 5—used a reflector of just 13 cm diameter to obtain a picture of the Sun. Even by the middle of this century, its sharpness had never been excelled by any other photograph. Janssen used a wet collodion plate, and the exposure time was 1/3000 second. Only later, when people photographed the Sun from the highest possible mountains, and then searched through thousands of pictures to find one during whose exposure the air had happened to be perfectly still, could Janssen's photograph be matched. Martin Schwarzschild wanted to get above all the atmospheric turbulence and photograph the Sun from a balloon.

In contrast to rockets, a balloon has the disadvantage that it is still inside the Earth's atmosphere, although it is so high that the unwanted effects are practically non-existent. This disadvantage is accompanied by the great advantage that observations may last for hours. The ideas behind project Stratoscope, as it was called, came from two Princeton astronomers, Martin

Schwarzschild and Lyman Spitzer. Schwarzschild soon took over the balloon project, and Spitzer immediately began working on plans that saw fruition 15 years later in the Copernicus satellite, one of the first astronomical observatories to orbit the Earth.

In the middle 1950s, funding for astronomical programmes was extremely low world-wide—even in the USA. Despite this, Schwarzschild succeeded in releasing the first balloon from Minnesota in September 1957. It reached a height of 25 kilometres. The gondola carried instruments to photograph the photosphere. Because Stratoscope flights were unmanned, control of the observation, such as directing the telescope towards the Sun and the exposure times, had to be pre-programmed. Once set into operation, the instruments were left to their own devices. All that Schwarzschild's team could do was to track the balloon from the ground with a telescope, follow it with a jeep, and hope that the instruments would withstand the shock of landing, wherever the wind might carry them.

When in October 1957, the first Soviet Sputnik peeped its way round the Earth and the USA slowly recovered from 'Sputnik Shock', the funding for Stratoscope became easier. Schwarzschild was able to fit a television camera to the balloon, from which it would transmit pictures down to the ground. He was thus able to see the image of the Sun obtained by the telescope on a screen, and to control the instruments remotely. The project needed the new, more generous funding, because the costs at that time for every single flight at the required height were at least 20 000 dollars, and a more costly flight might involve as much as a million dollars. Even more expensive than the flight itself were the repairs to the instrumentation that were required after each landing, which could not be controlled in any way.

In the meantime, numerous scientists had joined the team. Studies were not restricted to just the granulation. Robert Danielson (1931–1976) observed sunspots, and the penumbra in particular. John B. Rogerson investigated phenomena at the solar limb.

Stratoscope I—later followed by Stratoscope II, which observed the night sky—obtained the best photographs of the granulation up to that date. Janssen's photography showed granulation elements with diameters of between 800 and 1600 km, but the Stratoscope images showed that there were others, much smaller. It proved possible to make out features as small as 160 km across. It was found that the rising and falling clumps of gas were not irregular in shape. The boundaries of the hot gases, which appear brighter in the photographs, were not circular, but angular. Frequently, a more or less regular, hexagonal structure could be detected. This was not surprising. In the laboratory, liquids in a pan that is heated from beneath also exhibit hexagonal cells, with material rising in the centre and descending at the edges. It was comforting to know that phenomena that had long been familiar to us were duplicated on the surface of the Sun.

The balloon-borne observations were ideal preparation for working with space probes. Mind you, it was about the time of the Stratoscope observations that people first learned how to observe space from rockets. This was how the first images of the Sun in X-rays were obtained.

THE X-RAY SUN

X-rays cannot be collected by lenses or by the type of concave mirrors used in optical or infrared telescopes, so it is not easy to build an X-ray camera. Medical X-ray pictures are not produced inside a camera. They are, instead, the shadows that the organs being investigated (which are illuminated by a point source of X-rays), cast onto either a photographic plate or a fluorescent screen.

A total solar eclipse that occurred over the southern Pacific on 12 October 1958 helped to reveal a bit more about X-ray sources on the solar disk. At the beginning of the eclipse, six Nike-Asp rockets stood on the helicopter deck of the USS *Point Defiance*, ready to carry X-ray detectors above the Earth's atmosphere. The first rocket was launched so that it recorded the Sun's X-ray emission shortly before the beginning of totality, when a small crescent remained uncovered. This contained two areas of strong activity. The second launch captured X-rays from an even thinner crescent at the Sun's eastern limb, and by the time of the third, the Sun had been completely covered by the Moon. For the fourth, a narrow crescent had appeared at the western limb. Here there were two filaments. The all-important discovery obtained by this experiment was that X-rays are still detectable from the Sun when it is completely hidden behind the Moon.

It is by no means essential to use the Moon to scan the disk of the Sun and determine where its X-rays originate. Although lenses are not suitable, a pin-hole camera may be used, in fact it works as well for X-rays as for visible light. To put it more accurately: A pin-hole camera works as badly in X-rays as it does in visible light. At the time, no one had anything better.

On 19 April 1960, Herbert Friedman and his collaborators from the Naval Research Laboratory in Washington, DC, took a snapshot of the Sun in X-rays, using a pin-hole camera carried by a rocket. This historic picture is shown in Figure 12.1. Despite the shortcomings mentioned in the caption, it is possible to make out that in X-rays the Sun shows both bright spots and dark regions. It was only 13 years later that we discovered the reasons for this.

PROBLEMS WITH SKYLAB

The rocket-borne experiment during an eclipse that we have just described took place a year after the launch of Sputnik 1, the first artificial Earth satellite. The time was approaching when it would be possible to keep instruments in

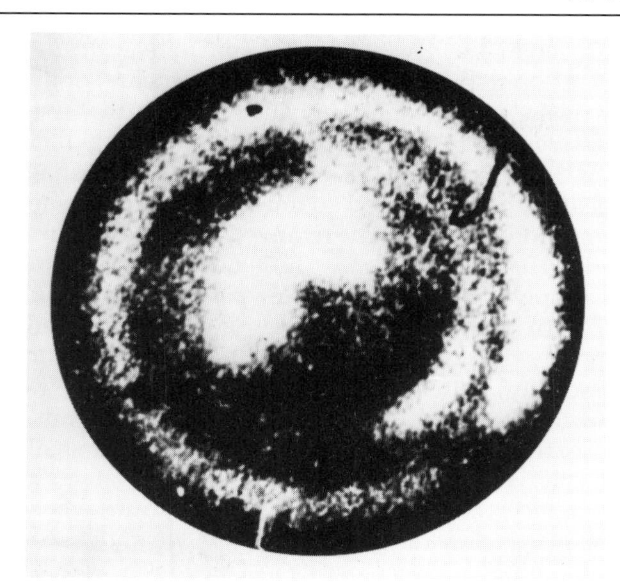

Figure 12.1. An X-ray image of the Sun, taken with a pinhole camera. Bright spots and darker areas may be seen on the Sun's disk, and it is also possible to see that the region immediately surrounding the Sun is also bright at X-ray wavelengths. Although the rocket was aligned with the Sun, it rotated on its axis during the flight, causing the bright spots on the Sun to be smeared out into arcs (Naval Research Laboratory, Washington DC)

space for as long as we liked, and be able to switch them on and off at will. The OSO—or **O**rbiting **S**olar **O**bservatory—satellites were built to observe the Sun. A total of eight satellites in this series were placed in orbit, the first of these in the early 1960s. Their telescopes were relatively small, and their results were overshadowed by events in the middle 1970s. True investigation of the Sun from space began on 14 May 1973 with Skylab, a manned orbiting space station.

Initially, it appeared as if it was going to be a failure. A solar panel, which was supposed to provide power for the space station, was torn off during launch. The second panel was still connected to the remnants of the first. Practically the whole electrical system was out of action. The crew were actually due to board the space station three days later, but during launch, part of the shield intended to protect the station from too high a level of solar radiation was also torn away. The temperature inside Skylab rose to about 50°C. In less than two weeks, a special shield was devised to reduce the amount of heating from the Sun. Once deployed this would cause the temperature inside the station to fall and become bearable. Taking the new protective shield and specially constructed tools with them, the astronauts were launched, and their capsule docked with the space station.

During the first repair, the new shield was spread over the damaged area of the outer hull. Despite this deployment, for the first few days the temperature in parts of the space station remained so high that the astronauts, Charles Conrad, Joseph Kerwin and Paul Weitz, frequently had to work in a temperature of 30°C. Some pieces of equipment, which had to fit predetermined apertures, had expanded so much because of the heat that they had to cool down before they could be removed from the overheated storage lockers where they had been stowed during launch. A further problem was the power supply. The remaining solar panel had jammed and could not be unfurled. An aluminium strap was holding it closed. The fault could be cleared only by working outside the space station; during an EVA (extravehicular activity). After the protective shield had been erected this second EVA had to be carried out. When any external repairs have to be carried out in space they require careful planning. At the Marshall Spaceflight Center in Huntsville, Alabama there is a giant tank of water. The buoyancy of the water counteracts the effect of gravity on astronauts, whose spacesuits act as diving dresses, so they can learn to work in weightless conditions. So that the astronauts could be given specific instructions how the repair to the solar panel could be carried out, replicas of the panel, its support structure and the offending aluminium strap were constructed in the tank, following the description given by Conrad and his crew. In the tank, astronaut Russell Schweickart used the same tools that the other astronauts had on board Skylab to free the panel. This enabled him to advise his colleagues in orbit, where Conrad and Kerwin spent three hours working outside the space station. The jammed panel unfolded and began to provide power. In the words of the American solar physicist Robert Noyes—'Then began nine months of operation, which among other things revolutionized the study of the Sun's corona.'

WHAT SKYLAB SAW

For the first time there was a large telescope in space. With it, among other things, it was possible to study the Sun's ultraviolet spectrum. Above all, this revealed helium lines that cannot be seen from the ground. A spectroheliograph obtained images of the Sun in the light of spectral lines with very short wavelengths. The calcium network, which may be detected in the light of calcium from the ground, was also found in light at shorter wavelengths, such as light from some of the oxygen lines. A particularly interesting feature was revealed by the image of the Sun in the light of a magnesium line. Light from this line originates at a great height, practically in the corona. This magnesium line did not show the network, which is a sign that the magnetic field's fine structure that is responsible for it does not extend out into the corona. In X-rays, however, other magnetic structures were clearly visible.

As we have already mentioned, there is no simple form of camera that can obtain pictures in X-rays. There is a trick, however, that may be used. X-rays are reflected from metal surfaces when they encounter them at a very shallow angle—at glancing incidence, as it is called. The physicist Hans Wolter (1911–1978) discovered this type of telescope in 1952. Since then it has been called a *Wolter telescope*, and may be used to obtain images of celestial objects in X-rays. The series of X-ray observations of the Sun carried out by Skylab showed that the magnetic fields at the surface extended out into the corona. The stronger the magnetic fields, the more powerful the coronal X-ray radiation.

In X-rays it is possible to see giant magnetic loops, whose feet are anchored to the solar surface, as well as open field lines that extend far out into space, with just one end fixed to the Sun. Some of the magnetic loops are more luminous than others, although the magnetic field strength is the same. It appears that parts of the corona where the lines are tightly bent are particularly hot and thus radiate more strongly.

That brings us to the question of why the solar corona has a temperature of two million degrees, whereas the photosphere beneath it is actually quite cold, with a temperature of only a few thousand degrees.

The discovery of the hot loops in the solar corona has brought an old idea, proposed by Ludwig Biermann (1907–1986) back into prominence. According to Biermann, sound waves propagate outwards from the zone where the granulation keeps the material in constant motion. These waves transport energy into the corona and are thus responsible for the high temperatures.

The X-ray images obtained from Skylab (Figure 12.2) show the coronal holes that we have already discussed (p. 141), as well as the bright X-ray points where the energy derived from the mutual annihilation of opposing magnetic fields is converted into heat. White-light pictures of the corona show rapid changes. Occasionally, giant bubbles rise into the corona. They escape from the Sun at velocities of around a thousand kilometres per second (!).

Skylab enabled us to see processes on the Sun that provided new avenues for coronal research. Actually that is not surprising. From Earth, it is possible to observe the corona without impediment only during a total solar eclipse. Coronagraphs show the corona only in the immediate vicinity of the solar disk. It is not possible to carry out any investigations of the corona at a distance of more than about one fiftieth of a solar radius from the edge of the disk. If we add together all the solar eclipses that have occurred in recent human history, we get a total duration of a few hours, during which the corona appeared in all its glory to observers on Earth. Thanks to Skylab, from which the corona could be observed nearly continuously, we now have a few thousand hours of observation time.

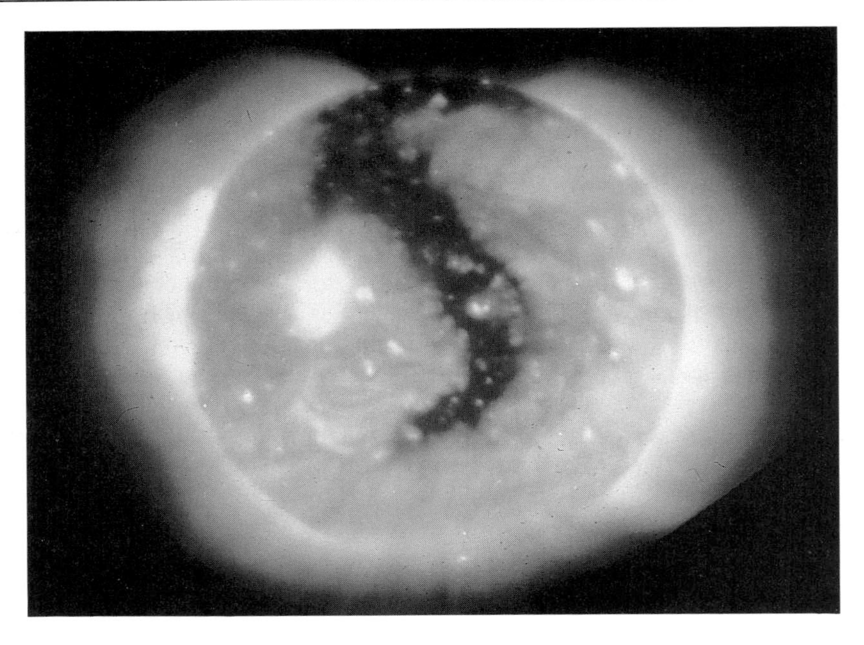

Figure 12.2. The Sun in X-rays. Outside the solar disk lies the luminous corona, through which we see the disk. The latter therefore appears bright because of the X-rays from the corona. There are also points on the surface of the Sun that are bright in X-rays. Particularly noticeable is the broad dark band lying across the Sun. This is a coronal hole, where coronal material has escaped into space, and which therefore does not emit any X-ray radiation

SKYLAB'S RETURN TO EARTH

The space station remained in orbit long after its scientific programme had come to an end. Over the course of time, nearly all satellites inevitably begin to lose height. They are subject to braking by traces of the Earth's atmosphere at the lowest portions of their orbits, which are generally just a few hundred kilometres in altitude. In July 1979, when the station was still orbiting at a height of 150 km, its decay became imminent. Initially it seemed as if parts would descend over the East Coast of the USA and Canada. The station was still under control, and the gyros on board were keeping Skylab oriented towards the Sun. If they were switched off, Skylab would begin to tumble, and its decay would accelerate. This offered the possibility of changing the point of impact so that it would occur in a sparsely inhabited area of the world, where there was little likelihood of damage. When it was feared that the station would not survive a further complete orbit, and would begin to tumble on its own, and thus be no longer under control, the gyroscopes were switched off by commands from Santiago in Chile and Madrid in Spain.

Skylab immediately began to tumble. Since its launch it had completed 34 981 orbits of the Earth, and now it began its final pass. The ground station on Bermuda gained contact with it for the last time. The solar panel was still working. Twenty minutes later, the tracking station on Ascension Island in the South Atlantic picked up the laboratory, which was slowly beginning to disintegrate. That was the last contact with Skylab.

As William Anderson, a pilot, was coming into land at the port of Perth in Western Australia on 11 July, he saw a bright blue light in the sky that turned to orange and later to red. Suddenly it burst into five individual luminous fragments. At 12:35 local time there was a long, bright trail in the sky over Perth. Sounds associated with the event were also heard. Over the next few days souvenir hunters searched the area for remnants of the wrecked observatory. It is estimated that about 20 tonnes of fragments reached the ground. Luckily it was an almost completely uninhabited area. That was the end of one of the most successful space missions, and NASA's first space station.

THE SOLAR MAXIMUM MISSION

During the nine months of Skylab activity in 1973/74, the Sun was shortly before sunspot minimum. The next maximum was expected to occur at the end of 1979 and the beginning of 1980. A satellite was launched on 14 February 1980 that was intended to monitor the Sun during the period of maximum activity. This unmanned satellite, known as the Solar Maximum Mission (SMM), had seven instruments on board, which were designed to study solar flares, in particular. The strength of solar radiation would also be monitored by SMM.

The satellite worked faultlessly for $9\frac{1}{2}$ months, before its pointing system failed. This was supposed to keep the instruments pointing at the correct spot on the Sun. Since then many other satellites have been launched that, after a time, have broken down, and had to be abandoned. This was not the case with SMM. Its orbital height of 600 km meant that it was within reach of the Space Shuttle. For this reason, the satellite had been built so that individual parts could be easily exchanged. Preparations for the repair took nearly three years. New tools were developed, and every manoeuvre was tested in the water tank under similar conditions to those prevailing in space. Finally everything was ready.

The Space Shuttle carried five astronauts into orbit in April 1984. After they had launched another satellite into orbit, Bob Crippen, captain of the *Challenger*, together with Astronaut Scobie (who was to be killed in the same spacecraft 21 months later), steered the craft alongside SMM.

NASA captured details of the manoeuvre on film, using on-board cameras. I had the opportunity to see a copy on video cassette. The damaged satellite appears, looking like a giant barrel glowing in the Sun, against the sky background. The two solar panels, which provide the satellite with power, look like wings spread on either side. Crippen brings the Shuttle to a distance of 90 m from the satellite. The two objects are now orbiting the Earth alongside one another, but the satellite is still spinning round its own axis. The next day, two astronauts leave the Shuttle. They have prepared for working outside the protected quarters by breathing an atmosphere of pure oxygen. They will breathe pure oxygen in their spacesuits during their work outside. Then one of them releases the manned manoeuvring unit (MMU) that will allow him to move freely around in space. The unit looks like a giant backpack or almost as if he had buckled on a dining chair with armrests, which actually contain the controls for the 12 nozzles through which the astronaut can blow nitrogen gas to propel him in any direction. On Earth, the MMU weighs nearly 150 kilogrammes, but under weightlessness it poses no problems for the astronaut, except that it moves correspondingly slowly. The trip in this nitrogen-powered armchair begins. In slow motion, the astronaut leaves the open cargo bay of the shuttle, heading towards SMM. After 10 minutes he has reached the satellite. He now has to stop SMM's rotation. To do so he has to couple the MMU to the side of the satellite, and then try, using the nitrogen jets, to cancel SMM's rotation. Once this has been done, the satellite may be brought into the cargo bay without damaging the protruding solar panels. During this operation the rest of the crew monitor the manoeuvres being carried out by the man in space, ready, if necessary, to rush to his rescue and take him on board again. But nothing dangerous happens.

Now the Shuttle closes to nine metres of the satellite. A specially constructed arm, controlled by Astronaut Nelson, grasps the observatory and carefully brings it down into a cradle inside the open cargo bay (Figure 12.3). SMM is secure, and the repair can begin. The orbital period of the Shuttle and SMM is 100 minutes. There is daylight for 60 minutes, and during the rest of the time they are inside the Earth's shadow. Floodlights illuminate the temporary workshop. Both astronauts have to change some of the components. They have to work carefully, despite the fact that, in orbit, the weight of the parts is unimportant. They actually have to move around masses that would amount to hundreds of kilogrammes on Earth. Once in motion, such components are not easy to bring to rest. In addition, the solar panels, which stretch out beyond both sides of the Space Shuttle, must not be damaged.

During a break in the work, all five astronauts are inside the Shuttle. The President of the United States is on the phone and NASA's film shows President Reagan at his desk in the White House. The five astronauts in the Space Shuttle are also visible, floating freely, with no shoes, just socks, while the President thanks them on behalf of the country. I cannot imagine that Americans often talk to their President in their socks.

Figure 12.3. A NASA artist's impression of the Solar Maximum Mission being repaired on board the Space Shuttle *Challenger*

Then the work outside begins again. Two astronauts switch components. During their work outside they wear safety lines and special, thick gloves to protect their hands from the surrounding vacuum. With them they have to plug in cables and turn screws. During the repair, two of the instruments on SMM are overhauled. Finally, it is ready for service again.

The grappling arm carefully lifts the satellite from the cargo bay, and the solar observatory is released. Solar Maximum Mission is once again in an independent orbit around the Earth.

In 1988, SMM was again in the news, when it became known that its measurements of gamma-ray radiation coming from space had been seriously affected by Soviet spy satellites. The cause of this was the nuclear reactors used to supply the orbiting spies with power. In doing so, positively charged particles escape from the reactors. These are positrons, which we discussed in Chapter 1. They are a form of antimatter and when one of them encounters a particle of ordinary matter, such as one of the outer panels of SMM, both particles annihilate one another in a flash of radiation. This radiation lies in the gamma-ray region, and thus affected the detectors on board SMM. Apart from this problem, SMM was one of the most successful scientific missions. Slowly, as the years passed, the satellite's orbit decayed deeper and deeper into the atmosphere. As late as November 1989, however, it was still delivering useful data.

On 2 December 1989 the 2268-kg observatory began its last orbit. Shortly afterwards it burnt up in the atmosphere over the Indian Ocean. During its nearly 10-year lifetime, the Solar Maximum Mission recorded 12 500 flares on the Sun. Apart from its main task, it also discovered several comets, made measurements of the famous supernova of February 1987 in the Large Magellanic Cloud in the southern hemisphere, and also carried out investigations of the Earth's ozone layer.

DOES THE SUN'S RADIATION VARY?

Our life on Earth depends on the Sun, and we know that its supply of nuclear energy will last for thousands of millions of years, and that there is no danger of life on Earth being subject to extreme heat or extreme cold. There remains the question, however, of whether the Sun may show smaller fluctuations in brightness, and that perhaps this may have occurred in what is known as the 'Little Ice Age', during the Maunder Minimum. It is not easy to monitor the Sun's output from the Earth's surface. Any increase or decrease in the amount of radiation reaching the Earth may be masked by variations in atmospheric transparency. This is why it is far better to measure the strength of solar radiation from space. The results of the measurements obtained by SMM during the first five months of its mission are shown in Figure 12.4. The Sun's output does actually vary. This variation, however, amounts to only about one tenth of 1 per cent of the total radiation. The variability is closely linked with the fraction of the solar disk covered in sunspots. When there are numerous spots on the side of the Sun that is turned towards us, the amount

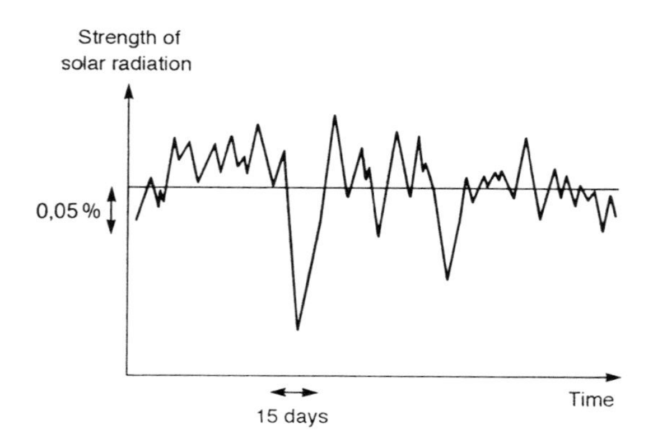

Figure 12.4. The variations in the strength of solar radiation, as measured by the Solar Maximum Mission, are less than one part per thousand

of radiation that we receive is slightly less. Because of the Sun's rotation, the variations in the solar radiation also exhibit the 27-day rotation period.

SPACELAB

At the end of the countdown, the liquid-fuel rockets ignited, followed by the solid-fuel boosters 6.6 seconds later. Everything appeared to be going to plan. It was 16:00 Universal Time, 11 o'clock in the morning at the Kennedy Spaceflight Center in Florida. The Space Shuttle *Columbia* slowly rose from the ground, moving straight up into the sky. A thrust of 30 million newtons accelerated the Shuttle, and 16 seconds into the flight it broke the sound barrier at a speed of around 1200 kilometres per hour. When the craft reached four times the speed of sound, the exhausted solid-fuel boosters were jettisoned. They slowly descended beneath parachutes into the sea, where recovery ships awaited them.

The Shuttle itself continued to climb, powered by the three main engines fed from a single giant tank. Six minutes later, these engines were also shut down and shortly afterwards the tank was jettisoned. By then, the altitude was 120 kilometres, and the tank re-entered the atmosphere at such a high speed that it burnt up. In the meantime the Shuttle had reached orbit. The date was 28 November 1983.

It was the ninth Space-Shuttle flight and the craft had the Spacelab laboratory on board. In the next 10 days, 7 hours and 47 minutes, the Shuttle and its cargo would complete 166 orbits of the Earth, before landing at an Air Force base in California at 23:47 Universal Time on 8 December. The mission was followed with particular interest in Germany, because for the first time a West German astronaut, Ulf Merbold, had been sent into space. Among the numerous experiments carried out during the flight, the strength of the Sun's radiation, particularly in the ultraviolet region, was re-measured.

It promised to be the beginning of a series of Space-Shuttle flights in which astronomical measurements, and particularly measurements of the Sun, would be carried out. Because of the *Challenger* disaster in 1986, however, the whole programme of Shuttle flights was disrupted.

Spaceflight has not only opened up windows into new regions of the spectrum, but it has also revealed exciting activity on the Sun that we would never have suspected in looking from the ground. It has also offered the possibility of directly investigating the material that the Sun ejects into space.

TWO GERMAN–AMERICAN ARTIFICIAL PLANETS

The instruments aboard Skylab and the later Spacelab studied the Sun from Earth orbit. The two Helios probes, on the other hand, flew directly towards the Sun. They were small artificial satellites, or rather, artificial planets.

Helios was a German–American collaborative project. In December 1974, a five-stage Titan-Centaur rocket was launched from Cape Canaveral. Its payload was the 371-kg Helios probe. This was boosted out of Earth orbit and placed onto a path that would take it around the Sun. At the time of injection into its transfer orbit, the probe was at the same distance as the Earth, i.e., 150 million kilometres, but it would later reach a distance of only 46 million kilometres from the Sun. That is closer than the planet Mercury, which orbits the Sun at an average distance of 58 million kilometres.

On 15 March 1975, Helios I reached the point in its orbit closest to the Sun for the first time. The radiation was 10 times as strong as near the Earth. Temperatures of 150°C prevailed on board. Despite this, nearly everything functioned without problems. Only one aerial, which had been intended to measure low-frequency waves in the plasma streaming away from the Sun, was found to be less sensitive than expected.

There were a total of 12 instruments on board. Seven of these had been constructed by groups from Germany, three by teams from the USA, and the remaining two were collaborative efforts. The data were radioed back to Earth, where the signals were captured either by NASA's radio telescopes, with 64-m diameter dishes, or by the 100-m dish of the Max-Planck Institute for Radio Astronomy at Effelsberg in the Eifel mountains. During the 190-day orbit, there were two periods when communications broke down: when the probe was either directly in front of, or behind, the Sun, whose radio emissions interfered with reception.

In January 1976, the sister probe Helios II was launched and injected into a similar orbit. It passed within 43.3 million kilometres of the Sun. The Helios probes were actually intended to fulfil their missions after flights of about three months. They were not designed for a longer lifetime. They continued to operate, however, and were used for a much longer time. After three years, Helios II encountered temperature problems; it eventually failed on 3 March 1980.

By the beginning of 1986, i.e., 12 years after its launch, the link with Helios I became difficult to maintain. The probe failed to respond to commands from Earth. Previously the orientation of Helios I had been maintained by the attitude-control jets which released gas on command from Earth, and this had kept the radio aerial pointing towards the Earth. Now this became impossible. Although most of the instruments were still working, the radio beam slowly drifted away from Earth, and no more signals were received from Helios I.

ULYSSES AND SOHO

The Helios probes had provided information about the surges of gas that the Sun expels into space, and which reach the Earth. However, the Earth's orbit and the orbits of satellites launched from it—such as those of the Helios probes—lie very close to the Sun's equatorial plane. As a result we know very little about the material that flows away from the regions towards the Sun's poles. This should be remedied by the Ulysses probe, which is a collaborative project involving the European Space Administration (ESA) and NASA.

A mission failure was nearly programmed into the probe before the launch. Luckily, shortly beforehand, it was discovered that a number of the chips that had been used were faulty, and they had to be changed.

The probe was launched in October 1990. In February 1992, it flew past Jupiter, where that planet's gravity hurled it back towards the centre of the Solar System and also out of the plane in which the Earth and all the other planets orbit the Sun. Its orbit then took it in a wide arc high above the poles of the Sun. It will pass over the South Pole in 1994, and over the North Pole a year later.

Scientists from 44 institutes have built instruments that are mounted on board Ulysses. Radio aerials will measure plasma waves, and both the numbers and the velocity of particles coming from the Sun will be recorded. Magnetometers will investigate the accompanying magnetic fields. Special detectors will record X-ray radiation from solar flares.

Later, in March 1995, the Solar and Heliospheric Observatory (SOHO) will be launched. This probe, which carries experiments built by teams of scientists from Finland, France, Germany, Great Britain, Switzerland and the USA, will monitor the Sun from a point 1.5 million kilometres from Earth, where the gravitational forces of the Sun and Earth are in balance. As well as the numerous instruments that will investigate the gases ejected by the Sun, together with the accompanying magnetic and electrical fields, there will also be instruments on board to record solar oscillations. Although it is possible to monitor the Sun for days on end from the South Pole (see p. 183), weather conditions prevent longer series of observations. SOHO will be able to monitor the Sun for much longer periods without interruption. Lying well inside the Earth's orbit, it will not even be affected by eclipses, because neither the Moon nor the Earth will interrupt its view of the Sun.

13

SUN AND EARTH

> With this increased activity the . . . strength of light-production may also easily gain in intensity, and thus also be capable of bestowing its gifts on our Earth in greater abundance and quality. Herschel, in an attempt to reach this conclusion, compared sunspots with the prices of corn that applied at the same time in England.
>
> Johann Heinrich Mädler (1794–1874)

The alternation of day and night, and of summer and winter, shows us that life on Earth depends on the Sun. Do we, on Earth, experience some effect of the changes undergone by the Sun over the course of the solar cycle?

Because the Sun is responsible for the striking difference in weather that we experience between January and July, this is bound to lead to the question of whether sunspots also affect our climate. Although they must have some effect on the weather on Earth as a whole, there cannot be any very strong connection, otherwise predictions about the weather, based on either the 11-year or 22-year cycles, would have been made a long time ago.

When we look at the measurements of the very slight variations in the strength of the Sun's radiation over time (Figure 12.4, p. 214), it is difficult to believe that these short-term variations of about 1 tenth of one per cent, which are caused by sunspots and disappear within 10 days, can have any marked effect on our climate. As yet, however, we have measured the intensity of the Sun's radiation accurately over just a short period of time. We do not know if it does systematically change over a solar cycle.

Although only recently have we known anything about the insignificant changes in the Sun's luminosity, records about our climate stretch back a long time. There are even records dating back long before Man learned to write.

THE MAN WHO HAD DOUBTS ABOUT THE MARTIAN CANALS

In Chapter 2 we mentioned some of the laymen who made important contributions to solar research, such as Schwabe the apothecary, and Spörer the school-teacher. There are similar examples in other fields of astronomy. Less well-known are the cases where an astronomer was particularly successful in another branch of science. Andrew Ellicot Douglass (1867–1962) is perhaps the most outstanding example.

Douglass was born in Vermont, and became interested in astronomy at an early age. At the age of 13, he was in charge of demonstrating the school telescope during an open day. Somewhat later, he described in the school magazine how to make telescope mirrors oneself, and how he had observed a solar eclipse. After finishing his studies—he chose astronomy, of course—he obtained a position at Harvard Observatory and worked at the out-station in Peru. Then he met Percival Lowell (1855–1916). The latter had built an observatory in Flagstaff, Arizona, devoted to the study of Mars. He was convinced that the fine lines that many observers thought they saw on the surface of Mars were actually artificial irrigation systems built by the Martian inhabitants. Lowell came from a rich, Boston business family and financed the observatory in Flagstaff from his own means. We now know that there are no Martian canals. At the time, however, the canals on Mars were the general topic of conversation, thanks to Lowell's propaganda, and much to the disgust of sceptical colleagues. Lowell offered Douglass a post at his observatory, which the latter accepted. Initially the matter fascinated him, but he soon began to wonder if the Martian canals were not basically an illusion. He began to take a critical look at the extent to which 'canals' were seen when illuminated spheres with smooth surfaces free from any markings were viewed through a telescope from a considerable distance. In 1901, this doubt about the reality of the Martian canals cost him his job.

Without a position, and with no prospects of being able to work as an astronomer again, Douglass tried politics. He was nominated by the Republican Party as a candidate for the judiciary. He won the vote, and at the same time, continued to work as a Spanish teacher. His teaching brought him back into academic circles, and he was soon given the task of setting up the Steward Observatory of the University of Arizona in Tucson. He was back in astronomy.

In the meantime, he had hit upon an idea that was to make him famous far outside the field of astronomy, namely the information that could be derived from the study of the annual growth rings in trees.

SUNSPOTS, TREE RINGS, AND PUEBLOS

When, at the end of 1901, Douglass travelled across northern Arizona and Utah, he saw how altitude above sea level and average rainfall influenced vegetation. The higher the region, the more numerous and the stronger the trees, which were more or less absent in the arid landscape of Arizona. The moisture was determined by the amount of evaporation from the oceans, and thus involved the Sun. Douglass concluded that trees were sensitive indicators of the strength of radiation from the Sun.

At around this time, people had begun to suggest that solar activity was responsible for the Asian monsoons. People also wanted to test whether sunspots decreased the average temperature on Earth. Douglass realised that the thickness of tree rings held information about past climatic conditions, and that this was an inexhaustible, and hitherto insufficiently used, source of data.

He began his investigation in a friend's wood store. The annual rings of felled trees were of varying thickness. Narrow ones succeeded wide ones, and it appeared as if the same sequence of narrow and wide rings was repeated in different tree trunks. Eventually he found two trees that revealed the same sequence. In one, however, the outermost 10 rings of the sequence were missing. Douglass concluded that this tree must have been felled 10 years earlier. A check on the dates confirmed this. So the problem was solved: The climate during any particular year determined the width of the ring for that year. When a year favoured growth, every tree laid down a wider ring than it did in a poor year.

Douglass now recognised the possibility of determining the ages of different pieces of wood by a method of comparison. If the same sequence of thick and thin rings was found, then the trees had been growing at the same time. To determine how many years elapsed between one tree being felled and another, all that had to be done was to count the number of additional rings, nearer to the bark, that appear in the younger piece of timber. Samples from still-living trees could thus be related to logs that had been felled a long time earlier, but which showed the same sequence in their outer portions as occurred in the centre of recent samples. Old logs were found in Indian pueblos that were still inhabited, and even older ones in ancient, uninhabited Anzani ruins in New Mexico. By the beginning of the 1930s, Douglass had assembled an unbroken sequence of overlapping tree-ring samples that covered 1900 years. The method that he developed of determining the age of timber helped to provide exact dating of archaeological finds in America.

Douglass began with the idea of finding indications of solar activity in tree rings, or at least being able to recognise climatic cycles. Samples from sequoias and Californian bristle-cone pines appeared to show 11-year cycles. Only during the period 1650 to 1740 were they not apparent. In February 1922, Douglass received a letter from Maunder, who pointed out the decline in the solar cycle over that period (see p. 29). To Douglass, this seemed to be an indication that evidence of solar activity could be found in tree rings. Unequivocal proof of this assumption nevertheless still remains to be found.

There is, however, no doubt that tree rings are influenced by climate. But do sunspots have anything to do with our climate

CLIMATE AND SUNSPOTS

For 200 years people have been searching for an 11-year cycle in the weather for which sunspots might be responsible. The quotation at the beginning of this chapter mentions William Herschel's attempt to detect a connection. Nothing came of it, and the astronomer, Mädler, from whom the quotation is taken, continues 'But it must have been very difficult to determine any influence, with the frequent wars, and the resulting changes in legislation, which must both have had an influence on corn prices ...' Other searches were undertaken, without success. At one time it seemed as if a 22-year cycle in the thickness of tree rings had been discovered. Closer study did not reveal any such cycle. Indeed, if trees from different parts of the globe are compared, differences in climate are certainly seen, but hardly any common properties indicating a worldwide influence, which is what would be expected from sunspots. Whatever influence sunspots have, they do not affect the climate globally, so that over the whole Earth tree rings are wider in one year than in another.

If sunspots do influence the overall weather, then their effects are not the same in different parts of the Earth. It is believed, for example, that measurements show more rainfall in the zone between 70 and 80 degrees North during sunspot maxima. On the other hand, rainfall measurements for the zone between 60 and 70 degrees North indicate more rain at sunspot minima than at maxima.

The news about annual tree rings and sunspots stimulated a whole range of similar investigations. People thought that they had found evidence of the sunspot cycle in harvests, in the price of furs, in the water level of Lake Victoria, and even in share prices. It is not easy to test such correlations. If variables that alter with time, whether they are the price of furs or water levels, are compared with sunspot numbers, one cannot confine one's attention to just the series of measurements that suggest such a connection. So, for example, a good return on rabbit skins during solar maximum should not only be present over the period between 1900 and 1940, but one should also find the same relationship over the next 40 years.

There can be no doubt about the carbon-14 abundances in annual tree rings that we discussed in Chapter 2. 'Radio-carbon', as it is called , is a definite indicator of solar activity. The more sunspots, the less ^{14}C is formed. Consequently, instead of searching for a relationship involving solar activity itself, about which there is less and less information the farther back in history we go, we can use the quantity of ^{14}C in tree rings, when we are trying to establish the connection between solar activity and past climate. Unfortunately, the results are contradictory. For example, in 1979, Hans Süß of the University of La Jolla in California wrote in the journal *Umschau* that there was a definite relationship between the climate in Europe and the ^{14}C-

content in tree rings during the Maunder Minimum and Spörer Minimum. In both periods it was somewhat colder in Europe. On the other hand, in 1980, Mince Stuiver of the University of Washington in Seattle, announced in the journal *Nature* that he could find no relationship between climate and the ^{14}C-content. He did not consider that the so-called 'Little Ice Age' in the seventeenth century had anything to do with the Maunder Minimum that occurred at around the same time.

In 1980, a sensation appeared to be in the offing. George Wilson of Adelaide in Australia attracted attention by stating that a certain 680-million-year-old sedimentary rock displayed layers of different thickness and colour. On average, every eleventh layer was darker, but he was also able to make out periodicities involving 22 and 90 layers. The layers in the Australian rock were thought to be deposits created on the floor of a lake by annual melt-water from Ice-Age glaciers. If the weather depended on the solar cycle, then one would also expect the amount of melt-water to vary with the solar cycle. The Australian rocks appeared to hold a record of the sunspot cycle millions of years earlier.

It was too good to be true. Since then the idea has fallen from favour. The erstwhile climate must have reacted extremely sensitively to sunspot numbers, far more so than nowadays. In fact the layers were probably not laid down yearly, but daily. They probably reflect the rhythm of the ebb and flow of the tides, together with the 14-day periodicity of spring tides. In addition, there is the monthly periodicity in the Moon's distance from the Earth. This combination of periodicities may be responsible for the Australian layering. We can envisage the rocks being laid down in a lagoon, at the mouth of a river that brought fine sand down from the land. When the flood-tide entered from the sea and encountered the current, the suspended material would be deposited sooner than when the river was able to flow unhindered through the lagoon into the sea. The tides could have caused the different layers in the rock. Sunspots were not responsible. So it is by no means easy to find any link between solar activity and terrestrial climate.

For decades people have been searching for relationships between processes occurring on the Sun and on Earth. Scientists like Douglass have spent their whole lifetimes searching for just such a definite link, with no notable success. So I was extremely surprised to discover a man—occasionally described in the German press as a 'space professor'—explaining everything in great detail to readers of a German television journal in January 1988, as if he had proved it all a long time ago: 'The gas explosions on the Sun hurl a vast quantity of atomic fragments into space . . . and muddle up our bio-rhythms. The delicate bio-states of plants are also affected (they no longer know whether they ought to bloom), and tiny animals such as termites, for example, change their feeding habits.' Not one of these supposed consequences would survive thorough scrutiny.

THE SUN'S ULTRAVIOLET LIGHT

When the Earth's orbit takes it slightly closer to the Sun in the northern winter, we hardly notice, on the Earth's surface, that the amount of incident radiation is about 7 per cent greater than it is in summer. The difference is completely submerged by the opposite effect caused by the change in season.

The insignificant reduction in the amount of radiation from the Sun when large sunspot groups appear on the visible portion of the disk, has a far smaller effect on the Earth. No effect of solar activity on terrestrial weather should actually be detectable.

The variations in solar radiation shown in Figure 12.4 relate to the whole spectrum. Measurements from space probes, however, have shown that short-wave radiation in the far ultraviolet and in X-rays varies much more with solar activity.

As we have seen in Chapter 3, ultraviolet light lies beyond the violet end of the visible spectrum. The K line of calcium, in which we can see the calcium network (see p. 88), lies at a wavelength of 3.9 ten-thousandths of a metre (390 nanometres). The form of ultraviolet radiation known as UV-A, which is harmless to us, begins about there, and extends to shorter wavelengths, down to to about 3.2 ten-thousandths of a metre (320 nanometres). At still shorter wavelengths this merges into the region of UV-B, which itself extends down to 2.8 ten-thousandths of a metre (280 nanometres). Then we come to the even shorter UV-C region, which is extremely dangerous to us, but which does not reach the surface of the Earth. Anyone who is sunbathing is exposed to just the A and B fractions of ultraviolet light. There is practically no tanning effect from UV-A. Sunburn comes from UV-B. The majority of this is blocked by the ozone layer in the atmosphere. This is just as well, because UV-B is injurious, giving rise to skin cancer. It is estimated that a reduction in the protective ozone layer amounting to just 1 per cent would cause a rise in human skin cancer of between 2 and 5 per cent. So worries about the propellants in spray cans and the refrigerants in refrigerators and freezers that destroy the ozone layer are well founded.

The amount of UV-B in the sunlight that reaches us on the ground depends on the altitude of the Sun above the horizon. The difference between when the Sun is high in the sky and when it is low is very considerable. Only when the Sun is more that three hand-widths above the horizon is it possible to get sunburnt. Anyone in southern Spain who suns themselves for one minute at midday in June receives as much tanning as they would in six hours at the same spot in December.

The fraction of ultraviolet light that is absorbed in the upper layers of our atmosphere affects the atmosphere's heat budget, and thus probably has some effect on our weather. Ultraviolet light may also be dangerous for artificial satellites orbiting the Earth.

HOW SUNSPOTS SHOOT DOWN SATELLITES

Luckily for us, most of the Sun's short-wave radiation in the UV-B and UV-C regions is blocked by the outer layers of the atmosphere, and thus prevented from reaching the surface and causing damage to humans, animals, and plants. The UV radiation warms the outermost layers of the atmosphere. In fact, the temperature rises at heights above 90 kilometres. At 200 km, air temperatures are around 500°C. This layer, the *thermosphere*, is highly susceptible to solar radiation. During periods of high solar activity, the UV-B and UV-C fractions increase markedly. As a result, when solar activity is high the thermosphere becomes hotter. In 1952 and 1962, both around sunspot minimum, the temperature was approximately 400°C, whereas in maximum years it was around 1100°C. When the thermosphere becomes hotter around sunspot maximum, it also becomes denser and extends farther out into space. Low-altitude satellites feel the effects of this. If the thermosphere suddenly expands as far as their orbits, friction increases, the satellites are braked, and they decay sooner than expected. This is why Skylab, which continued to orbit the Earth, unmanned, after all its planned tasks had been completed, experienced greater drag than expected during the sunspot maximum of 1979/80. The space station burned up early. The Sun had, so to speak, shot down the very artificial satellite built to study it.

Solar activity influences the activities of spaceflight authorities. The Hubble Space Telescope (HST) is an American–European collaborative project in which some 1.3 thousand million dollars had been invested at the time of its launch. From orbit, it was expected to provide astronomers on Earth with the best possible conditions for observing the sky. It was scheduled to be carried into orbit by the Space Shuttle in June 1986, when it would have been commissioned during the last sunspot minimum. The *Challenger* disaster delayed the launch by years. In fact it was eventually launched on 10 April 1990. The telescope began its life in orbit in the middle of a sunspot maximum. It had to be launched into a more distant orbit than originally planned, to avoid it being too rapidly braked by the Earth's atmosphere that had expanded as a result of solar activity.

Because of the great significance of the degree of solar activity for spaceflight, which has to be planned years in advance, it is essential to know how active a sunspot maximum will be. Will the relative sunspot number at the next maximum be less than 100, or more than 200? To what extent can we predict the strength of the next maximum from the behaviour during minimum? Currently, the planners of spaceflight missions do not have any reliable methods of predicting what will happen.

It is not just the stronger ultraviolet radiation from the Sun that heats and expands the Earth's atmosphere. The particles ejected from the Sun at extremely high velocities also affect the outmost layers of air, and imperil astronauts as well as satellites.

ASTRONAUTS IN DANGER

The maiden flight of the Space Shuttle *Columbia* in 1981 had to be delayed by 48 hours for technical reasons. If the astronauts John Young and Robert Crippen had taken off on 10 April, as planned, they would have been subjected to one of the strongest showers of radiation in recent years. If *Columbia*'s orbit had taken it over the poles, where the electrically charged particles from the Sun tend to precipitate, and the astronauts had needed to work outside the Shuttle, as they did in repairing Skylab (see p. 206) and SMM (see p. 211), the high-energy protons from the Sun would have pierced the aluminium shielding layers of their spacesuits. Young and Crippen would have been in mortal danger. Astronauts who leave their spacecraft always run the risk of being exposed to an overdose of dangerous particle radiation. This also applied to the Apollo astronauts who walked on the surface of the Moon. This was why investigations of the dangers from particle radiation began in the 1960s. NASA set up a warning system, because showers of particles may be predicted, at least in the short term. The majority originate in flares—at least that was what was thought until recently. We shall return to this point later on p. 229.

If a flare is observed on the Sun, in most cases a shower of particles reaches the Earth about 50 hours later. Monitoring the Sun for flares therefore offers the possibility of predicting showers. It is not just the day-to-day work of a space station that is influenced by eruptions on the Sun. The consequences that particles from the Sun have on Earth were shown by the event that nearly caught the Space Shuttle *Columbia* in April 1981.

THE FLARE OF APRIL 1981

On 2 April, the point on the Sun that was the site of later events showed just a single spot. Sunspots are accompanied by magnetic fields, which move around following the motion of the material into which they are frozen. The magnetic fields become stretched and twisted, which stores energy in the fields. If, later, field lines annihilate one another, this energy is released as a flare.

Up to 7 April the spot did not change. Then it produced a flare, but not a particularly strong one. The next day 16 new spots appeared, and five additional flares were observed. On 9 April there were 29 spots. Eight more flares were seen. The magnetic field detected over the area that day was extremely complicated. A vast number of tiny areas of north and south polarity were lying next to one another. Five small spots of various polarities surrounded the large spot that had been observed since 2 April. Only 10 per cent of spot groups show such a complex field. Solar experts knew that this was just the sort of configuration that could give rise to a major flare.

It came on 10 April, 43 hours before the launch of *Columbia*. The solar surface brightened over an area of two million square kilometres and remained in this state for $3\frac{1}{2}$ hours. During this time the amount of energy released was equivalent to about 100 thousand million atomic bombs of the size of that devastated Hiroshima. Such a vast amount of energy would satisfy the current energy consumption (from all sources, including coal, oil and gas), of the whole of Germany for about a million years, or of the United States for about one sixth of that period. The energy must have come from the magnetic field, because the sunspots did not change during the event.

In the first 10 minutes a shockwave raced outwards through the corona. Fifty-eight hours later, a stream of electrically charged particles encountered the Earth's magnetic field and caused a magnetic storm. The strength and direction of the Earth's field oscillated wildly. The changeable magnetic field induced such extremely strong currents in power cables that the Canadian power supply network broke down. Electrons crashed into the upper atmosphere and raised the temperature at an altitude of 260 km by 1500°C to 2500°C. Regular measurements of the temperature at that altitude had been obtained for 15 years, but such a large rise had never been recorded. The Earth's atmosphere expanded and caused a sharp drop in the orbital height of the Space Shuttle *Columbia*, which had been launched in the meantime. The astronauts on board had to fire the engines to correct their orbit.

Although the eruption of April 1981 was particularly important because of the Space Shuttle flight taking place at the same time, other events in the past have provided a source of excitement. The flare that Carrington observed in 1859 was followed by a magnetic storm that affected the whole Earth. Normal telegraph links were broken, and the oscillations of the magnetic field produced such a strong current in a cable that it could be used to send telegraph messages, even though it was not connected to any battery. Aurorae lit up the nights in both hemispheres, even as far as the tropics. Magnetic needles oscillated wildly all on their own.

Although the event in April 1981 was the strongest of that particular cycle (No. 20), the eruption of August 1972 in Cycle 19 holds the record in the popularity stakes. The particles emitted by a flare caused current surges in the American telephone network, and produced such massive changes in the voltages in power cables that circuit breakers automatically disconnected the power supply in many states of the USA and Canada. Ships in the St Lawrence Seaway were left without radio contact. Aurorae were seen over areas of the Earth where they had never been seen before.

PROTONS FROM THE SUN

The April 1981 event was caused by a stream of high-speed protons. The fastest of these reached the Earth an hour after the flare. This means that they left the Sun with a velocity of at least 40 000 km/s. The majority arrived two days later, having thus travelled at about 900 km/s. On 10 April, the number of particles arriving at the top of the atmosphere amounted to 300 per second per square centimetre. The protons were not alone, because their positive charges dragged lighter electrons after them. All these charged particles did not reach us along straight lines, but along the magnetic field lines that stretch out of the corona into space. Because of the Sun's rotation, these lines are spiral in shape (Figure 13.1). The charged particles reaching the Earth have travelled along curved paths, so the only particles that can reach us are those that originate in flares observed on the western half of the disk. Particles from flares in the eastern half miss the Earth.

This is an important point for anyone trying to predict proton storms caused by flares. Recently, however, a famous astronomer has revealed a new aspect of interplanetary weather forecasting.

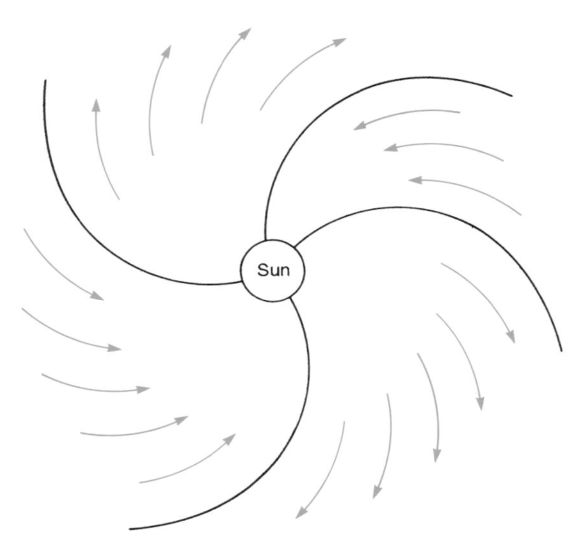

Figure 13.1. As the solar wind escapes into space, it carries magnetic fields of different polarity along with it. Different sectors (here indicated by black lines) are observed, in which the field lines point either outwards or inwards, relative to the Sun. This diagram is just a schematic representation of the true properties of the magnetic fields

PROTON STORMS AND FLICKERING RADIO GALAXIES

At Cambridge in England there is an unusual radio telescope. There is no giant radio dish to be moved and pointed at a particular point in the sky. Instead, 2048 individual dipole elements are spread across an area of 1.8 hectares and interconnected to form an aerial array. By means of appropriate switching the direction of the array's beam may be altered. The aerial then receives radiation from a point at a specific height above the horizon, instead of just from the zenith. The Earth's daily rotation carries the sky across the radio beam every 24 hours. In this manner, it is possible to direct the beam at any given point in the northern sky at some time during the day, and measure the strength of the radiation that it is emitting.

Using this array, Anthony Hewish, professor of astronomy at the Cavendish Laboratory in Cambridge, and his collaborators have given a new boost to the question of proton storms from the Sun. They were, however, not using their telescope to investigate the Sun, but were instead studying radio emissions from stellar systems known as radio galaxies, whose radiation has taken millions of years to reach us. In studying these distant objects, Hewish came to the conclusion that not all proton storms originate in flares. Showers of protons reach the Earth that cannot be ascribed to any flare. An example is the storm of 27 to 29 August 1978. It was so strong that an auroral display was seen even in Santa Fe in New Mexico, at a latitude of just 35°N. This shower did not originate in any flare. The network set up by NASA to warn of proton storms was completely surprised. No one would have been able to warn any astronauts working outside their spacecraft.

What do the distant stellar systems that Hewish and his collaborators were studying have to do with events on the Sun? The protons and electrons that the latter ejects form a plasma. Some portions of the stream of plasma ejected by the Sun are denser than others, and thus contain more particles per cubic centimetre. The condensations may be thought of as 'clouds' in the ejected plasma. Radio waves are very sensitive to the properties of the medium through which they are passing. Distant radio sources 'twinkle' when seen through the solar plasma.

We know what happens when light passes through the Earth's atmosphere. Rays of light from stars are repeatedly deviated from their paths by layers of different density in the atmosphere, which are in constant motion. This is why stars twinkle. As with starlight passing through the atmosphere, the radio emission from galaxies is continuously disturbed by irregularities in the stream of plasma emitted by the Sun, causing the strength of the radio signal to vary. Radio sources 'twinkle' in the solar plasma. If we investigate the radio luminosity of distant galaxies we can deduce something about the stream of plasma emitted by the Sun.

Hewish's team investigated the strength of the *scintillation*, as this variation is called, of 2500 galaxies daily for two years. In doing so they repeatedly found areas of the sky where the scintillation was particularly strong. These regions of strong scintillation changed their positions on the sky. On one day a galaxy might show little scintillation, on the next began to show changes in its radio luminosity, and on the third exhibited extremely strong scintillation. It was found that these regions of strong scintillation were actually moving. They all originated in the Sun and moved away from it across the sky.

Where precisely do these clouds of plasma from the Sun originate? According to the Cambridge astronomers, they were not related to flares, but arose from coronal holes. It had been known for a long time that streams of particles were emitted by coronal holes, but it had been believed that they gave rise merely to the relatively tame *solar wind*, which we shall discuss shortly. Hewish believed that he had now shown that showers of high-energy particles could occasionally be emitted by coronal holes, and that these produced aurorae on Earth, were responsible for the loss of radio communications and power-supply failures, and could endanger astronauts.

THE SOLAR WIND

It was known that solar eruptions ejected material and that this could reach the Earth's orbit. No one suspected, however, that the Sun was constantly losing material to space, even when no eruptions were observed on its surface. The clues that eventually led to the discovery of the solar wind were shown by celestial objects that, at first sight, had nothing to do with the Sun. These were comets, which occasionally spread long tails across the sky. They rise and set with the stars, and yet are unlike the fixed stars in that, over days or months, they move across the constellations.

It had been known for a long time that the tails of comets always point away from the Sun, as if they were repelled by it (Figure 13.2). The question of what force might be responsible for keeping cometary tails pointing away from the Sun could not be answered until Ludwig Biermann made a close study of

Figure 13.2. The tail of Halley's Comet in 1986. The solar wind is responsible for the fact that cometary tails always point away from the Sun (photo. European Southern Observatory, Chile)

the direction of the tail of Comet Whipple-Fetke, which appeared in 1942. It had been noted that tails did not point precisely away from the Sun. Biermann realised that this could be explained if the comet was moving in a flow of gas itself streaming away from the Sun. The difference in the direction of the tail and a line pointing directly away from the Sun was a result of the comet's own motion relative to the flow of gas. Biermann concluded that even when the Sun was quiet and no spots were visible, there was still a continuous flow of gas through interplanetary space. Cometary tails fluttered in this stream of material like flags in the wind. He also succeeded in estimating the velocity of the ejected material. To a certain extent, it could be said that Biermann discovered the solar wind from his desk. It was only later that the existence of the solar wind from the quiet Sun was established by space probes.

We now know that the solar wind consists principally of electrons and hydrogen ions. The solar wind has a velocity of about 400 km/s, and sometimes considerably more. At the Earth's distance it has a density of about 10 particles per cubic centimetre. As we have already seen in Chapter 8, it flows out of coronal holes.

At that time Biermann was working at Göttingen. It was there that another scientist had made an important contribution to the problem of the material ejected from the Sun. The Chair of Geophysics at Göttingen had a long tradition. In 1833, the great mathematician and astronomer Carl Friedrich Gauss (1777–1855) had established there the first laboratory that regularly monitored the Earth's magnetic field. Every variation in the direction and strength of the field was recorded. In the 1930s, Julius Bartels (1899–1964) noted that the variations indicated a periodicity of about 27 days. If the magnetic needles waved about very strongly on one particular day, then on average they would probably do the same 27 days later. A period of 27 days makes one think immediately of one solar rotation. Bartels therefore concluded that there were regions on the Sun that—for whatever reason—caused the Earth's magnetic field to fluctuate when they were turned towards the Earth. When, a few days later, the Sun's rotation caused them to point in another direction, the Earth's magnetic field became quiet. But 27 days later, the rotation again caused them to lie in a position from which they could affect the Earth, and the magnetic field was disturbed once more. Bartels called these hypothetical regions on the Sun, *M-regions*, the 'M' standing for 'magnetic'. No one knew what the regions were on the Sun, and some of Bartels' colleagues maintained that the 'M' stood for 'mysterious'.

Since we have had X-ray images of the Sun, we know the answer: Bartels' mysterious regions and coronal holes are one and the same. The solar wind blows out of them. The two Helios probes (p. 215) investigated them, and determined what sorts of particles were ejected from the Sun, in which directions and at what velocities. They established that the solar-wind material flows smoothly out into space along the magnetic field lines extending from the Sun, as one might expect with a plasma. That was not particularly surprising. What was amazing, however, was the fact that the magnetic field lines of the solar corona should reach far out into interplanetary space without being 'nipped in two' by the finite electrical conductivity of the plasma.

Another surprise was that the solar wind contained helium atoms, which had lost only one of their two electrons. By their very nature, an atomic nucleus of helium is normally accompanied by two electrons. At high temperatures these are removed in collisions with other atoms. In the hot solar corona, with its temperature of a million degrees, nearly all the helium atoms have lost both electrons. Yet the solar wind contains many helium atoms that have regained an electron. Do they, perhaps, not come directly from the corona? Has somehow cool material from the layers beneath the corona succeeded in escaping into space, without being heated in the hot shell surrounding the Sun? Have they been protected inside larger clumps of material, and thus survived their journey through the Sun's hot corona?

There are hardly any questions in solar research that—as soon as one tries to answer them—do not raise further questions. We are closer to the Sun than to any other star, and we are able to see a whole range of phenomena on its surface. It makes us realise how little we know about other stars, which, even in the largest telescopes, remain just points of light.

Although the Sun's proximity may not help us to understand all the processes that are occurring within it, it may at least help us to make practical use of the Sun.

14

TAPPING THE SUN

> If we were to add to the current false price for energy production the costs of all its detrimental effects, such as damage to forests, climatic change, and the management of nuclear wastes, i.e., if we were to draw up the true economic balance, then solar energy would no longer be so much dearer.
>
> Ludwig Bölkow

When the Roman fleet showed up outside Syracuse, it was attacked using solar energy. Allegedly, the Greek physicist and mathematician Archimedes (287–212 BC) had large mirrors built, with which he concentrated the Sun's rays on individual ships, until they burst into flames. Whoever devised this incredible story of the application of solar energy in ancient times obviously had military uses in mind. Anyone who wants to make use of solar energy nowadays thinks of peaceful applications. One would be forced to have international collaboration for the necessary construction and use. It is possible, however, that many smaller solar power stations, scattered over the country, might fulfil the aims better.

So far not much has happened. Current solar power stations are nearly all pilot models. Many of them are still confined to the planners' drawing-boards. Indeed many of the projects resemble science-fiction. When Herr Meyer recently found out about the plans of Dr Peter Glaser, the research director of a firm in Massachusetts, who is proposing a solar power station in Earth orbit, the idea fascinated him so much that it occurred in one of his dreams.

HERR MEYER ABOARD THE SOLAR POWER STATION

They were all very friendly. First of all, Herr Meyer had been welcomed by the director of the station. Now a young engineer was guiding him through the corridors. It was not easy to move under the weightless conditions. Without his special velcro shoes, Herr Meyer would have lost his footing at every step.

'Tomorrow I go back to Earth', the young man suddenly said. 'Two months working up here, then a month off. That's how we live.'

As they passed a porthole, Herr Meyer looked out at the vast surface, glinting in the sunlight.

'Our solar collectors cover a circular area three kilometres across.'

Herr Meyer already knew that the station was in geostationary orbit. It orbited the Earth once in 24 hours from West to East. As a result the same side of the globe was visible all the time. As seen from the Earth, the satellite remained at the same point in the sky.

'With our output of five gigawatts, we are comparable with a large atomic power station', Herr Meyer's companion explained. 'The collector surface is covered with gallium-arsenide cells, which convert sunlight into electrical current. Our microwave transmitter radiates the energy to an array of aerials in southern Italy. There the microwaves are converted back into electrical energy. Although energy is lost in the form of unusable heat, we are able to feed about half of the energy that we capture out here into the European power grid. That's an extremely good return!'

Herr Meyer remembered the introduction from the station director:

'We have about 350 men and women on board all the time, to keep the station operating. The most important thing is that we keep our microwave beam trained on the array in Italy. That beam is the one dangerous thing about our operation. However, our computers ensure that we do not stray off target. A laser beam from the Earth is used to guide our pointing. If we miss the ground antenna it goes out, and our transmitter array is immediately switched off, to ensure that the microwaves cause no damage. At present we are not really certain whether the radiation, spread over an area of 10–13 square kilometres, would actually cause any harm. Birds that nest around the receiving antenna appear to thrive. We don't have roast pigeons falling out of the sky!'

'The great advantage of capturing solar energy in orbit is that we are able to trap sunlight day and night. Only between September 1st and October 15th, and again between March 1st and April 15th, i.e., around the equinoxes, does the Earth hide the Sun for a maximum of 72 minutes each day. During these solar eclipses we are temporarily unable to trap any energy.'

When Herr Meyer looked out of the porthole again, he saw the Earth's disk rising above the collector array. Europe and Africa were surrounded by blue oceans. Automatically, Herr Meyer's eyes tried to pick out the Italian boot. Clouds obscured Calabria and Sicily.

'The microwaves penetrate the thickest cloud cover', the young engineer explained. When he saw how impressed Herr Meyer was, he added:

'This solves all Mankind's energy problems. Fossil fuels such as coal, oil and natural gas are no longer required. No more carbon dioxide is being added to the Earth's atmosphere. The danger of the greenhouse effect has been banished. We can do without nuclear power stations, which, at least since the Chernobyl incident, have made everyone feel uneasy. Mankind has been freed from a nightmare.

'It all began when governments on Earth began, one after the other, to invest more in the development of solar energy than in nuclear energy.'

When Herr Meyer awoke, the engineer's last words were still ringing in his head, and he was happy to know that the Earth's energy problems had been solved once and for all. Then he remembered that this had only happened when governments had provided greater resources for research into solar energy than into nuclear power. He thought of his own government and realised that it had all been a dream.

Herr Meyer's dream was not mere fantasy. The American scientist Peter E. Glaser developed plans for a solar power station in orbit as early as 1968. Thirteen years later, the National Reseach Council of the National Academy of Sciences checked every detail of the project, and came to the conclusion: 'that no funds should be allocated for further development of this project in the next ten years'. Such a station was certainly 'technically feasible, if costs were disregarded', it stated. This relegated Peter Glaser's amazing project to the cupboard. It will be a very long time before we obtain power from orbit, especially as there are more realistic potential methods of using solar energy. Unfortunately, these projects smack of science fiction. Our politicians are hesitant about taking them up, even though they would be considerably cheaper than orbiting space stations. An orbiting solar power station will be a long time coming. Although we are able to establish and maintain manned orbiting stations, these all have relatively low orbits, just a few hundred kilometres above the ground. A satellite in a geostationary orbit, which orbits the Earth in the same time as the latter takes for one rotation—and therefore appears to remain in the same spot on the sky to an observer on the ground—is at an altitude of 36 000 kilometres. It is impossible to achieve such an altitude with a Shuttle flight. Despite this, people are thinking seriously about obtaining solar energy from orbit. Only recently I saw a similar study by the German aerospace firm of Messerschmidt-Bölkow-Blohm. It considers using a laser beam to transfer the energy captured by an unmanned satellite down to Earth.

THE SUN, THE STAR THAT GIVES US LIFE

The first person to recognise that all life on Earth originates with the Sun, without which no rivers would flow, and no breath of air would stir, was neither a biologist, nor a physicist, nor an astronomer. He was a king. Or rather, a pharaoh:

Thou didst create the world according to thy desire.
Whilst thou wert alone;
All men, cattle, and wild beasts,
Whatever is on earth, going upon (its) feet,
And what is on high, flying with its wings.

This was the prayer offered up by Akhenaton, the husband of Nefertiti and father-in-law of Tutankhamun, to his god, the Sun. He was also the first to accept the concept of a single god. The priests of his day did not let him get away with it. After his death he was described as a heretic. Works of art from his reign were destroyed and his name was hacked off monuments with chisels. His belief in the omnipotence of the Sun fell into obscurity. It was only in the last century that people again realised that all life owes its existence to the Sun.

The Sun's light and warmth enable life to exist on Earth. The Sun will support us for thousands of millions of years to come. The dangers that threaten us today are of our own making, not ones hanging over us from space.

Apart from the solar energy that enables plants to grow and provides humans and animals with food, we have other essential energy requirements. It all began when people first used fire, and used the solar energy stored in wood to warm themselves and to prepare their food. Later coal and oil were added, again forms of solar energy that were captured and stored millions and millions of years ago. Even the energy that we obtain from hydroelectric schemes originates in the Sun, which evaporates water from the oceans, and carries it up into the hills and mountains. Fossil fuels such as coal and oil have many disadvantages. The most serious, however, is that when they are burned they release carbon dioxide into the atmosphere.

THE GREENHOUSE EFFECT

It is carbon dioxide in the air that makes the Earth habitable. If there were none of this gas in our atmosphere, the average temperature of our planet would be below freezing. Vast areas that are now inhabited would be buried beneath ice. The fact that we have a habitable climate is thanks to one of the properties of carbon dioxide. It does not allow any long-wave heat radiation to escape into space. At ground level, most of the Sun's energy falls in the visible region, because the atmosphere is transparent to those wavelengths. This warms the ground. At night, the energy that has been gained is radiated away again. If all of this energy were radiated freely away to space, the Earth's mean temperature would be significantly lower. Luckily, the night-time radiation does not escape so easily, because the carbon dioxide in the air restricts the escape of heat from the ground. It prevents cooling, because it is not transparent to long-wave heat radiation. The greater the quantity of carbon dioxide in the air, the greater this effect, which is known as the *greenhouse effect*. When the Sun shines on a greenhouse, the rays penetrate the glass, which is transparent to visible light. But when heated objects inside the greenhouse emit heat, they are only able to do so at long wavelengths, to which the glass is not transparent. As a result, the incoming solar energy warms the interior of the greenhouse. It is only at a higher temperature that the greenhouse is able to emit as much energy as it gains, because although the glass lets light in, it is reluctant to let heat out. This is why greenhouses are warmer than their surroundings. Similar considerations apply to our atmosphere. The Earth's average temperature is about 18°C higher than it would be if no carbon dioxide were present. The atmosphere of Venus consists almost entirely of carbon dioxide, and the greenhouse effect has raised its temperature to nearly 500°C. If we enrich our atmosphere in carbon dioxide by burning fossil fuels, we enhance the greenhouse effect. The average temperature of our planet will increase, the polar icecaps will melt, and sea level will rise.

Warnings of the greenhouse effect are no mere ghostly writing on the wall, as may be seen from the data collected by the Nimbus 5 and 6 satellites, which have revealed that in the last 15 years the Arctic Ocean's ice cover has shrunk. Perhaps it will only be when we have each encountered definite evidence for ourselves that we will take the warnings seriously and switch to solar energy.

The amount of solar energy falling on the area of a country such as Germany—which may be taken as an example of an industrialised, urban society—is about 100 times as much as that consumed. Unfortunately, using the flow of energy that falls, free, from the sky every day is by no means simple. To get a feeling for the amount of energy the Earth receives from the Sun, and Mankind's overall requirements, we need to be able to measure and compare amounts of energy.

FROM A KILOWATT-HOUR TO A TERAWATT-YEAR

Energy is often measured in kilowatt-hours. We all know approximately how much that is. At present, electricity supply companies in the majority of industrialised nations charge around 10.5–11 cents per kilowatt-hour for that amount. A typical light bulb is 100 watts. If we keep 10 such bulbs alight for an hour, we have used one kilowatt-hour. Note that a kilowatt is not a unit of energy. It is rather a unit of energy that is used or provided *per second*. A physicist would say that a watt is a unit of power. The Sun radiates so much power that, expressed in watts, it would be a 27-figure number. An area of one square metre exposed to the Sun at the distance of the Earth—

such as a satellite's solar panel—receives slightly more than one kilowatt. That would suffice to keep our 10 light bulbs alight as long as the solar panel was illuminated. In an hour, therefore, the area would collect slightly more than one kilowatt-hour. When we confine our discussion to just a few square metres of surface and ordinary, household light bulbs, watts, kilowatts, and kilowatt-hours are appropriate units in which to express power and energy. But we are about to discuss far larger amounts of energy and the kilowatt-hour is far too small. We need additional units.

A billion watts (one American 'trillion', or a million million watts) is called a *terawatt*. That is an unimaginable amount of power. If we express the Sun's luminosity in terawatts, we obtain a value that is only a 15-figure number. If a body radiates for a year at a power of one terawatt, it requires an amount of energy known as a *terawatt-year*. This is the unit of energy that planners use when they make projections of Mankind's energy requirements between now and the next century. The whole population of the Earth used about 2.4 terawatt-years in 1950. Twenty-five years later, this had already risen to eight terawatt-years per year.

The amount of energy involved in a terawatt-year is inconceivable. Even comparisons do not really bring much enlightenment. Nevertheless I shall try to give an example: The earliest traces of life on Earth are 3.5 thousand million years old. They consist of single-celled organisms, similar to present-day blue-green algae. If such a blue-green algal cell had switched on a 100-watt bulb and the latter had remained burning until now, it would have been shining when the primordial continents were created, and the first molluscs arose. It would have shone for the interminable time when tree-ferns were the major plant forms. The first insects would have fluttered around it. The first reptiles and snakes would have seen its light. It would have acted as a beacon for the first birds and the great saurians. At night, it would have attracted mammals, when they first appeared on the planet, and it would have illuminated the nights for the early humans, such as the Neandertals. When people began to paint bison on the walls of caves, they would probably have included an evocation of the everlasting light from our bulb. If it had shone for all those thousands of millions of years with a constant 100 watts, it would have used one third of a terawatt-year. At our current electricity prices it would, however, have meant that we would be presented with a bill for 300 thousand million dollars.

An area equal to that of Germany (357 000 square kilometres), which is slightly more than half the area of Texas, receives about 40 terawatt-years of energy from the Sun, every year, absolutely free. That would be enough for the whole world's population. Unfortunately, as yet, that incident energy cannot be used effectively. We can see, however, that Mankind could continue to live from solar energy, as it has done in the past—provided we can manage to use the incoming radiation from the Sun.

A GREENHOUSE ON THE ROOF

There are a number of possible ways of converting solar radiation to usable energy. The simplest way is by using it to warm water. Sunshine on a roof warms water which flows through suitable pipes spread across the surface. This is not enough for one to dispense with the electricity supply, but it can help to reduce the monthly electricity bill, because even at moderately high latitudes such as those in Germany or the northern United States and southern Canada, between 60 and 70 per cent of hot water requirements may be met in this way.

There is another possibility: The whole area of the side of the roof turned towards the Sun is covered by a glass-topped chamber. This creates a greenhouse effect. Water, in thin tubes, flows through the warm air in the chamber. The effect may be amplified by making the bottom of the chamber of a reflective material, shaped to concentrate the reflected light onto the tubes of water. Even when skies are grey, enough infrared radiation passes through the clouds to warm the tubes of water inside their roof-top greenhouse. Currently, these attempts have an air of improvisation and of 'do-it-yourself' methods, and it does not seem as if we will have many households supported by solar energy in the near future. An installation pays for itself only after 20 years or so, because at our latitudes we cannot completely do without any electricity, nor without oil, gas or coal for heating during the winter. How many people are prepared to install some system in their own private homes that will only become financially profitable after a couple of decades? Industry has never invested much money in full-scale mass-production of home systems, and so they remain expensive.

MIRRORS IN THE DESERT

Since the beginning of the 1980s solar power stations have been operating in sunny California. Numerous reflecting surfaces in the form of long troughs collect light and bring it to a focus—or more precisely, a focal line running along the trough. In Dagett, there are two arrays that have been operating since 1984 and 1985, where the focus consists of a pipe through which oil flows. The sunlight that has been concentrated on the pipe heats the oil, the latter flows into a tank, or heat-exchanger, where water circulates through a second set of pipes. The oil brings the water to boiling point, and the steam drives a turbine. In nearby Barstow, there is a similar array, where the water is turned to steam directly at the focus. Since 1982 its peak output has been 10 megawatts, and at one stage the solar power stations in the Mojave desert had a total output of 195 megawatts.

In Sicily, at the foot of Mount Etna, is Eurelios. The name is a compound of 'Europe' and 'Helios', the Greek word for the Sun. The project was financed by the European Community. A total of 189 nearly flat mirrors follow the daily motion of the Sun across the sky from East to West. Seventy of these reflectors were made in France, and 112 in Germany. The total collecting surface amounts to 35 000 square metres, i.e., 3.5 hectares. The carefully mounted mirrors reflect the sunlight onto the top of a 55-m high tower, which Italy contributed. The individual mirrors act together like an enormous single reflector, concentrating the heat at the top of the tower. There the temperature reaches 800°C, and water is turned into steam, which drives a turbine. The output is one megawatt. This, the first European experimental installation, has been in operation since 1981. It was the world's first solar power station. It is not far from the place where Archimedes is reputed to have set fire to the Roman fleet. He would undoubtedly have been delighted to see these 187 mirrors.

It is possible that turning solar energy first into heat and then into energy is not the best method. Perhaps we should be thinking of obtaining electricity directly from sunlight.

LIGHT INTO ELECTRICITY

The method of using hot water or oil is not required if we use materials known as *semiconductors*, whose properties fall between those of good conductors, such as metals, and those of insulators. These materials are used in the diodes and transistors that have made modern electronics possible, and which are found in everything from pocket calculators to television satellites. The material that seems almost made for converting sunlight into electricity is widely present on Earth. It is the element silicon, which forms a significant proportion of all rocks. As a result, it ought to be 'dirt-cheap', but unfortunately an involved process is required to turn rock into usable solar cells of silicon, so the latter are expensive. Silicon cells provide electricity for satellites in space. A few are already used on roofs to convert solar energy into electricity. In future, silicon-based solar cells may provide the whole of Mankind with electricity from solar energy.

I do not intend to discuss the complicated processes that occur within a solar cell in great detail here, but I will describe just a few of their properties.

Two, differently prepared, thin layers of silicon are in contact. As with all materials, the atoms have positively charged nuclei, which are accompanied by the same number of negatively charged electrons, so that the whole atom is electrically neutral. As soon as light falls on the silicon junction, however, electrons are knocked out of the atoms. These electrons are able to move freely between the atoms. The atoms themselves, because they lack electrons, are positively charged. Each requires one or two electrons to compensate for the positive charge on the nucleus. Electrons from neighbouring atoms are able to jump into the spaces left by the lost electrons. Those spaces are thus filled, but new ones have been created in neighbouring atoms. These are occupied by yet more electrons which, for their part, leave gaps even farther afield. So both electrons and the empty sites move in opposite directions through the thin layers of silicon. The electrons are negatively charged particles, so because there is nothing to fully compensate for the positive charge on the nuclei, the sites in the silicon atoms—physicists call them 'holes'—behave like positive charges. Like the electrons they move through the silicon. In principle, electrons and holes may jump from one layer to the other and back again. We have discovered, however, how to control the flow of electrons and holes.

The two different layers of silicon are prepared in advance. Slight traces of other materials, each another type of element, are introduced. The atoms of these additional materials give the two layers specific electrical properties. For example, electrons may be able to flow freely from one layer to the other, say from top to bottom, but not in the reverse direction. Electrical fields along the contact surface stop them if they attempt to move in one particular direction. The holes, on the other hand, move without hindrance from bottom to top, but again motion in the opposite direction is blocked. So the top layer becomes enriched in holes, and the lower with electrons. The top layer becomes positively charged and the lower negatively. Just like the top and bottom of a button-cell battery, electricity may now be drawn from the double layer of silicon. So we have obtained electricity from light. So all we need to do is produce silicon junctions, set them in the sunlight, and Mankind's electricity problems are solved! Unfortunately, it is not so simple.

The silicon junctions are not cheap. They have a limited lifetime, and the voltage that they produce is half a volt. Anyone who wants to obtain electricity from the Sun will have to wrestle with these three difficulties.

The problem of the low voltage is the simplest to solve. Just as one connects two 1.5-volt batteries in series to obtain 3 volts, so one can connect several silicon cells together in series. If we want 12 volts, like that of a car battery, we need 24 cells. In fact, when we draw any current the voltage drops slightly, so we actually need 30 cells.

Anyone who launches satellites into space can usually afford to mount solar cells on the solar panels. The Canadian communications satellite that was launched in 1976 carried 25 000 cells, each with an area of four square centimetres. They provided more than a kilowatt of power.

In recent years the prices of solar cells have dropped steeply. The cost of cells to provide one watt in the middle of the 1970s was about 50 US dollars, but 10 years later it was $5. In 1988, in view of the lifetime of cells, the price of one kilowatt-hour was estimated in the USA as being between 30 and 40 cents. By the end of 1989, a solar kilowatt-hour was 15 cents—still considerably above the current American price of electricity (10.5–11 cents per kilowatt-hour). Silicon is probably not the only material that may be used to convert sunlight to electricity. This may cause the price of cells to drop even farther in future.

Because solar cells have lifetimes of only about 20 years, they are able to trap energy over this period of time. In any case energy is required to produce them. Metallic silicon is produced from quartz and carbon in an arc furnace, and this must be purified. It is turned into vapour, allowed to condense, and revaporised, perhaps several times. All that costs energy. If more energy is required to manufacture a solar cell than it can produce in its working lifetime, then solar cells are not capable of solving Mankind's energy problems. The procedure just described is, however, only one of many that are able to produce usable sheets of silicon. For some time people have been searching for energy-saving methods of manufacture. A review of the subject published in 1988 in the German science journal *Physics in our Time* showed that cells produced by cheaper methods recover their costs within seven to eight years. A method has been found, however, that uses so little energy in the production process, that the cells recoup their costs in the first six months of use. For the rest of their lifetime they deliver an energy surplus.

MIRROR ARRAYS AND SOLAR POWER STATIONS

Are silicon cells and their remarkable properties only suitable for providing power to pocket calculators, as well as a few home-owners, who are prepared to burden themselves with high investment costs to obtain environmentally friendly energy? Are they being tried in just a few solar power stations, that will never be able to make a serious contribution to our energy requirements? Could solar cells solve Mankind's energy problems?

Although earlier estimates suggested a doubling of world energy requirements, energy-saving measures have refuted those earlier gloomy predictions. The energy consumption in developing countries is indeed still rising, but in industrialised countries it began to drop from the beginning of the 1970s, thanks to energy-saving programmes. The savings among the rich countries may compensate for the rise in requirements for the poorer countries, so for the present we may envisage a yearly world energy requirement of about nine terawatt-years.

Let us begin with Germany as an example of an advanced, industrial country. In western industrialised countries the energy consumption per head of population in 1975 was about 54 000 kilowatt-hours. Although we are normally used to measuring electrical energy in kilowatt-hours, the figure just quoted includes the total energy requirements for heating and transport. For the German population of 79 million, that amounted to about one half terawatt-year for 1975.

Frequently nuclear energy is touted as an alternative to the use of fossil fuels with their accompanying greenhouse effect. Currently, however, in the USA only 2.3 per cent of oil, 5.5 per cent of natural gas, and 81 per cent of coal is converted into electricity. Anyone wanting to use nuclear energy to meet just electricity requirements would still be spewing carbon dioxide into the air from nearly all the oil and natural gas that is consumed. Warding off the greenhouse effect means that nuclear energy will not only have to be brought to houses for use as heating, which ought to be relatively simple, but also made available in cars' fuel tanks. Let us assume that this problem can also be solved—solar energy faces the same difficulty in replacing the Tiger in the Tank with the Sun in the Tank (see p. 246). How many nuclear power stations would be required, if we wanted to meet the total energy requirements of Germany? The nuclear power station at Brockdorf has an output of 1335 megawatts. In a year it therefore produces 12 thousand million kilowatt-hours, which is 1.3 thousandths of a terawatt-year. The whole of Germany's energy requirements could therefore be met by 364 nuclear power stations the size of Brockdorf. Only if we are prepared to live with nuclear power stations, with the nearest rather less than 18 kilometres away, can we dispense with solar power (Figure 14.1). If we could make use of the waste heat from the

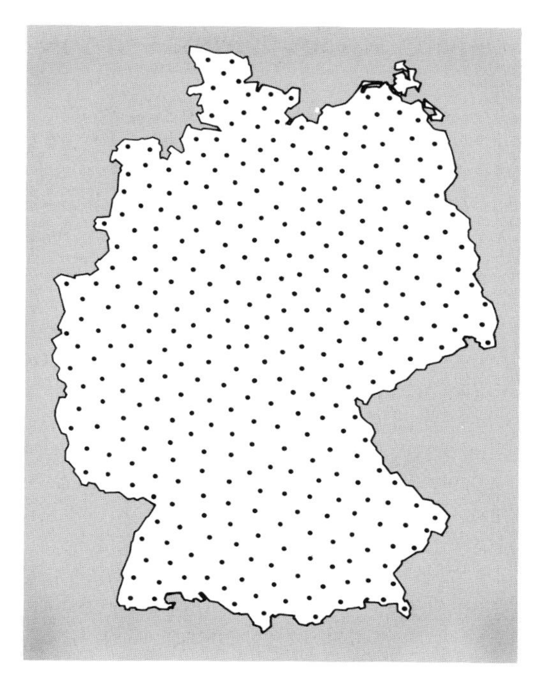

Figure 14.1 If we wanted to use nuclear energy to meet the whole of Germany's energy requirements, including heating and transport, we would have to cover the country with a network of nuclear power stations the size of the one at Brockdorf

power stations, the number would be somewhat less. It would still be far too large. Worldwide we would have to build 7000 such power stations to ward off the greenhouse effect.

What does it look like for solar energy? How large an area must we cover with silicon to meet Germany's total private and industrial energy requirements? If we assume that each cell is able to convert about 10 per cent of the incident solar energy into electricity, and also if we take into account that at mid-northern latitudes the Sun's rays fall at an angle, rather than vertically—when the Sun shines at all—we arrive at a figure of some eight million hectares. If this area were in the form of a circle, its diameter would be 340 kilometres. If the power station were sited in a region with more favourable climatic conditions, such as in the Sahara, where there is no agriculture, nor woods or nature reserves, one would probably find that a circular area 100 km across would suffice to meet Germany's power requirements. A circular area 700 km across in the Sahara would meet the whole world's energy needs. This again assumes that solar cells convert 10 per cent of the energy that they capture from the Sun into electricity. Figure 14.2 shows the area of the Sahara that would need to be covered with solar cells. Naturally, no one would

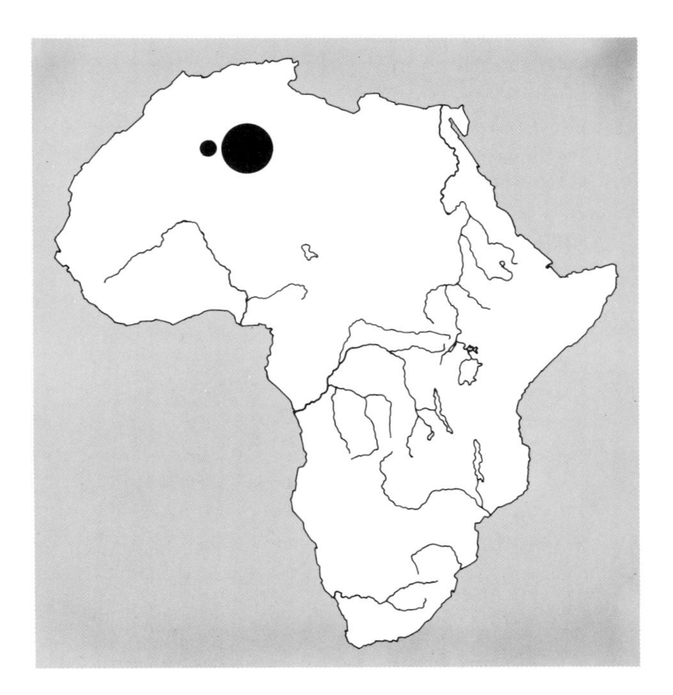

Figure 14.2. If we tried to meet Germany's energy requirements with solar energy from the Sahara, we would have to cover an area, equivalent to the smaller of the two circles, with solar cells. The larger would be sufficient to supply the whole world's energy requirements

construct a single power station to supply the whole world with electricity. There are many desert areas, and solar power stations with smaller collecting areas could be spread more or less evenly across the globe.

At first sight it might seem as if there are no energy problems, if we want to dispense with fossil fuels, which would otherwise damage our atmosphere, and if we simultaneously want to adopt the use of solar cells, rather than nuclear power. Unfortunately, there still remains a whole host of unsolved problems. Conversion from one form of energy to another does provide new jobs, but a large number of other employment opportunities are lost. In the long term, this may not involve problems for the economy. When electricity came to be used for street lighting, the job of lamp-lighter also disappeared. But it is easy to imagine the hardship that would follow if, for example, we wanted to run down the coal industry over a short term. A miner who loses his job today, cannot run around a solar cell array in the Sahara tomorrow in a jeep, changing solar cells that have failed. Undoubtedly the most important task facing politicians in the future will be the conversion to new methods of energy production, and ensuring that our social structure is able to cope.

Quite apart from all the other difficult problems that solar energy raises, there is also the dilemma facing our choice of sites to be used. Even when we have mastered all the other problems, are able to produce cheap solar cells, and have built power stations in the desert, there remains the fact that we are not producing the energy where it is required. It has to be transported all round the world. Linking the whole world with high-tension power lines and deep-sea cables would involve enormous additional costs, which could not be offset by any reduction in price of solar cells. People have been trying to think of a way out of this dilemma for a long time.

SOLAR ENERGY IN A BOTTLE

Do you remember the school experiment—or perhaps you tried it yourself as a child—where two wires are immersed in water, each connected to one pole of a an ordinary torch battery? Small bubbles climbed up each wire, more on one than on the other. The bubbles on one side were hydrogen, the others oxygen. The electrical current had dissociated the molecules of water, which consist of oxygen and hydrogen. Whereas at room temperature the water molecules form a liquid, the two fractions are both gases, which rise to the top of the water. This process uses energy from the battery to split water molecules. We know that oxygen and hydrogen may be recombined. If we collect the two types of gas in a container and hold a match to the opening, the two gases recombine with a small explosion to give water. The electrical energy that was required to split the water atoms is released in the form of heat. The highly explosive oxygen and hydrogen mixture is known as *oxyhydrogen*.

If we used the electricity captured in the Sahara to split water, then we could compress the hydrogen and oxygen and fill bottles of gas, which could then be exported—naturally separately, because then the gases are harmless. Only when they were brought together would there be a danger of explosion. In fact, only the hydrogen needs to be exported, because oxygen is found throughout the atmosphere.

So we can imagine that in future there will be installations in the deserts, which will use solar energy to extract hydrogen from water, and export it all over the world. It may be used in suitable boilers for heating purposes, or in engines. Hydrogen-powered cars and aeroplanes have yet to be mass-produced, but they are, in principle, attainable with our current degree of technology. Trial models have already been built.

Do we need to first convert solar energy into electricity, to obtain hydrogen? Is it not possible to allow the solar energy to act directly on water? In the outermost layers of the thick atmosphere of Venus, the intense solar radiation has dissociated all the water vapour molecules. The hydrogen has escaped into space, and the oxygen has remained. Might it not be possible to do the same on Earth? We have not yet got that far. We are able to split water molecules with light, but that is successful only if we add more energy. The same applies to a chemical reaction, where light helps to combine carbon dioxide molecules with water molecules to give methyl alcohol and oxygen. This process is also possible in principle, but so far we have not succeeded in carrying it out without adding energy from outside. Currently, we cannot gain any energy from the process.

Undoubtedly we ought to invest a considerable amount of money in the development of new technologies. Nuclear power remains the favourite with many politicians. Up to 1987, Germany had spent about 30 thousand million DM (roughly 18 thousand million US dollars) in developing nuclear energy. The amount devoted to solar-energy research has been minimal in comparison. In February 1986 a joint project between Saudi Arabia and Germany was begun. A total sum of less than 50 million DM (30 million US dollars) was set aside for this.

At some time in the future, Mankind will be forced to realise that its energy needs will have to be met directly from solar energy. No one knows when that will be, perhaps only after more Chernobyl-type disasters, or when the carbon dioxide from the burning of coal and gas has increased the greenhouse effect to such an extent that the first coastal towns are inundated by water from the melting polar ice-caps. It is possible that before then many will have to die, perhaps more than in the last World War, before people learn to live sensibly with our environment. Let us hope that it will not be too late.

APPENDICES

A SUNSPOTS AND A PAIR OF BINOCULARS

It is possible to use the projection method (Figure 2.2) to observe sunspots without any particular equipment. All that is required is a pair of binoculars and a piece of white card. If you intend to follow the suggestions that I am about to offer, please remember, *never, ever* look at the Sun through any telescope or pair of binoculars!

Hold the binoculars sideways in front of you, so that the objectives (the larger lenses) are pointing towards the Sun. The eyepieces (the smaller lenses) are then pointing away from the Sun. Now try to alter the alignment of the binoculars so that they point straight at the Sun, and the light passes through the objectives and out at the other end. You need the image from only one of the objectives, so cover the other one with a paper cap. Now you are only a couple of steps away from success.

First you need to intercept the light that passes through the binoculars with the piece of card. It should be held at right-angles to the Sun's rays. The light passing through the binoculars will produce a bright spot on the card, and this is the image of the Sun. It is probably still very fuzzy, so that no details are visible. By changing the distance between the binoculars and the card, you can obtain a sharp-edged circular disk, the brightness of which decreases slightly towards the edge.

If the disk shows dark spots that move with the image of the Sun if you shake the binoculars slightly, then they are sunspots. If, however, the spots remain stationary when you shake the binoculars, but the image moves about, they are caused by the binoculars' optics and are nothing to do with the Sun.

You will probably find that the sunlight that passes outside the binoculars still falls on the screen and badly affects the image. This light may be blocked by using a piece of paper with two openings to fit over the two eyepieces, so that only light that passes *through* one half of the binoculars reaches the card.

If after all this, you still see no sunspots, you are just unlucky and happen to be observing when the Sun is free from spots. However, this is fairly unlikely, because the next sunspot minimum will fall in the second half of the 1990s. It may be that the image of the Sun is not completely sharp.

It may be that you have actually tried to use a pair of opera glasses, which are constructed on the same principle as a Galilean telescope. It consists of a positive lens, such as people have to wear when they are long-sighted, and a negative lens (the type required for short-sight). With a Galilean telescope (or opera glasses) it is possible to obtain an image of the Sun, but it is small, and the screen has to be held at a considerable distance. Binoculars and astronomical telescopes are built according to the principles of Keplerian telescopes (with two positive lenses). With them it is easier to project a large image onto a screen. Scheiner (possibly) and Hevelius (certainly) used a Keplerian telescope (Figure 2.2). Kepler himself, as we have seen, used Regensburg cathedral. As far as the optics are concerned, Kepler had the advantage over us, because, in contrast to hand-held binoculars, the Regensburg cathedral did not wobble about. In fact, if you can arrange for the binoculars to be held rigidly, so that the projected image on the screen remains perfectly steady, you can immediately see greater detail in the image. Figure A.1 shows how you can make a slightly more complicated arrangement for projecting the Sun. Apart from the binoculars, you need a tripod, a mirror, and a cardboard box.

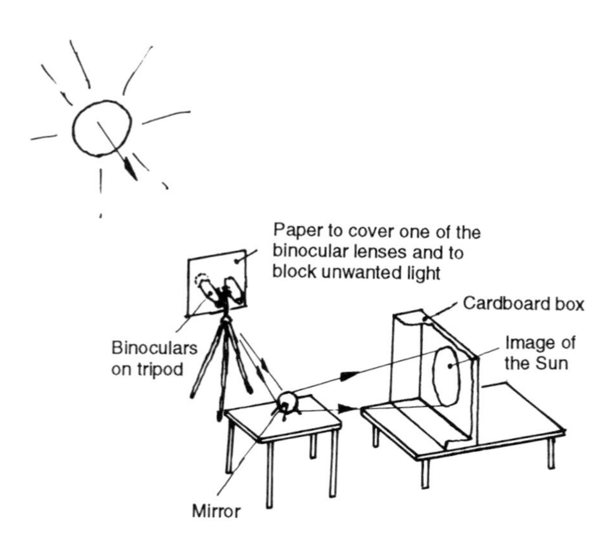

Figure A.1. A sketch drawn by the biology student Antje Wichelhaus to show visitors to the Solingen public observatory how she observed sunspots using a pair of binoculars

B THE MEGAPHONE INSIDE THE SUN

It is relatively easy to understand how, for sound waves, the pattern of oscillations at the surface provides information about the deep interior of the Sun.

Sound waves produced at the surface penetrate into the body of the Sun. What happens as they move down into the lower layers? Let us imagine that we have installed a large loudspeaker at some point on the Sun's surface, and that this emits a specific note. In addition, the loudspeaker does not radiate the sound equally in all directions, but in one specific direction, rather like the way a megaphone channels sound wherever one happens to point it.

This does not actually happen on the Sun, of course. If a particular point, e.g., one on the surface, oscillates at a particular frequency, waves spread out in all directions. This nearly always leads to an overall pattern of oscillations like that shown in Figure 10.7, which is not surprising, but which does not immediately tell us anything about what is happening. In what follows we will consider just a portion of the wave, and one that is emitted in one specific direction. Oscillations that involve the whole of the body concerned may be regarded as consisting of an infinite number of individual, directional beams, similar to the one emitted by our megaphone. So let us look more closely at the behaviour of the 'beam of sound' emitted by our megaphone, which points downward at an angle into the body of the Sun.

The sound-beam consists of a series of denser and more rarefied regions, following one another at the speed of sound. In the outermost layers of the Sun that amounts to five kilometres per second. In the Earth's atmosphere, the thunder following a flash of lightning propagates at a speed of only 350 metres per second. The speed of sound in the Sun is much higher. This is because of the higher temperature. The hotter a gas, the faster sound waves propagate through it. At the centre of the Sun, where it is far hotter than at the surface, and the temperatures are around 10 million degrees, a sound wave covers a distance of 400 kilometres every second!

In our beam of sound, the points of maximum and minimum density form more or less flat surfaces, as indicated in Figure B.1 by the two different grey tones. The dark surfaces correspond, so to speak, to the 'wave-crests' and the lighter ones to the 'wave-troughs'. The diagram also shows the beam of sound, with its crests and troughs, as seen from the side. Let us follow this beam of sound as it penetrates into the solar interior.

Figure B.1 shows that as the sound-wave progresses downwards the right-hand edge of each crest (or trough) penetrates slightly deeper into the solar interior than the left-hand edge. It therefore encounters a region where the speed of sound is slightly higher. As a result, it moves faster than the left-hand edge. The wave-crest does not maintain its original alignment; it turns slightly. Because the beam of sound propagates at right-angles to the plane of the wave-crests, it also changes direction. The angle that it makes with the

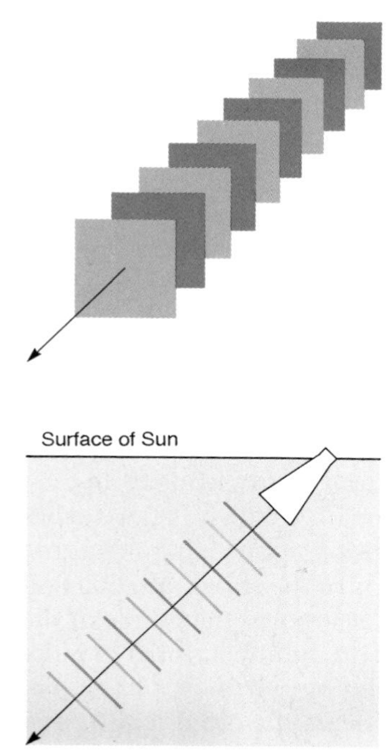

Figure B.1. Top: When a beam of sound penetrates into the Sun, the denser (dark grey) and rarified (light grey) regions succeed one another as almost plane surfaces. Bottom: A beam of sound emitted by a megaphone, as seen from the side. The dark and light grey perpendicular lines represent the wavefronts shown in the upper diagram

surface gradually becomes shallower and shallower. Eventually it becomes horizontal, and the planes containing the wave-crests and troughs are at right-angles to the surface. But the deeper portions of the wave-fronts continue to move faster than those at lesser depths. The beam of sound turns back towards the surface from which it came. After a while it reaches the surface. The beam of sound that we sent down into the Sun penetrated the deeper layers, but was then reflected back to the surface. What happens then? The Sun's surface acts just like the free end of the rope that we discussed in Figure 10.4, and reflects the wave back down again. This is shown schematically in Figure B.2. At regular intervals the beam of sound penetrates down into the interior, is turned back towards the surface, from which it is reflected back down into the lower layers.

The Sun is round. Any news that we shouted through our megaphone—which, for the sake of argument, we assume we pointed towards the South—returns after a few hours, but this time from the opposite direction (the North). In the meantime it will have been turned back towards the surface, and reflected back from the latter, a large number of times. It will have passed both the South and North Poles (Figure B.3).

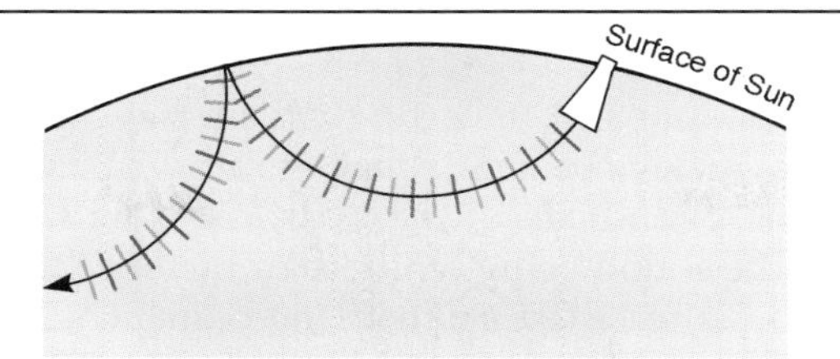

Figure B.2. A beam of sound-waves sent down into the interior of the Sun at an angle is bent back upwards. Those portions of the wave-fronts that penetrate deeper into the Sun move faster, because the speed of sound inside the Sun increases with depth. This is why the sound is bent upwards. At the surface it is reflected back down into the interior again

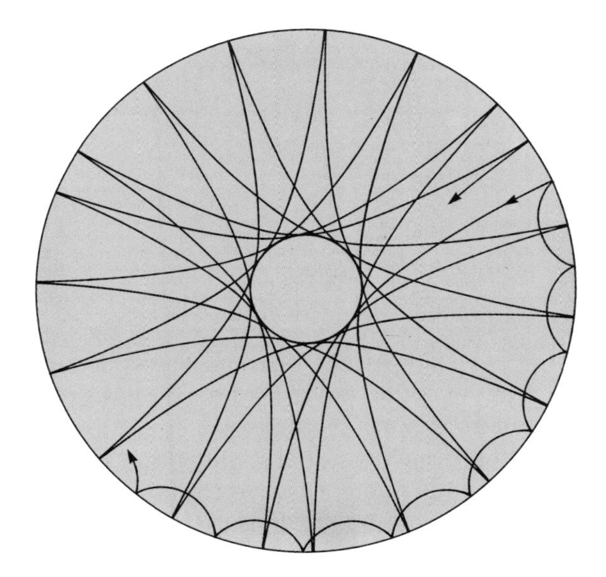

Figure B.3. The more acute the angle of a beam of sound, the deeper it penetrates into the solar interior. The combination of curvature within the interior, and reflection at the surface cause the sound-wave to migrate around the Sun and eventually return to its starting-point

After having completed a circuit of the Sun, the sound-wave may not return to its exact starting-point. We can, however, alter the direction of our megaphone slightly, pointing it down at a more acute angle, for example. The arc that the waves would describe would be more acute, would penetrate deeper into the interior, and would return to the surface at slightly different points. By changing the angle of the megaphone, we can ensure that the echo returns to the starting-point. At a very shallow angle, the wave may have reflected back into the interior only once (say), whereas at a much steeper angle it might be reflected perhaps 10 times. Anyway, let us assume that the echo returns to the starting-point. If our megaphone emits the same note for hours on end, this note and the echo that has travelled around the Sun combine. The echo may have become considerably weaker, but its frequency has not changed. The periodicity of the wave-crests and troughs is exactly the same as it is for the sound being emitted for the first time.

The returning echo is reflected again at the point where the megaphone is situated, and then propagates alongside the newly formed sound-wave. This raises the question of whether the old wave-crests and troughs coincide with the new ones. If so, the waves that have already travelled a long way will reinforce the new ones. Or will the old wave-crests coincide with the new trough? The echo would then weaken the newly emitted signal. It is immediately obvious that the answer depends on whether we can 'fit' a complete number of wave-crests and troughs into the length of the path that the echo has taken, so that after one circuit the wave-crests coincide. If this is not the case, we can slightly alter the pitch of the note, and thus the distance between the crests. In this way we can always ensure that the new beam of sound and the old echo reinforce one another.

A double megaphone

If such a set of waves travelling towards the South are reinforced by their echoes, then the same would apply to an identical set travelling towards the North. The two systems of waves running round the Sun would form a standing wave. Let us therefore take a second megaphone, with the same pitch, that emits a beam of sound down into the Sun at precisely the same angle, but this time pointing North (Figure B.4).

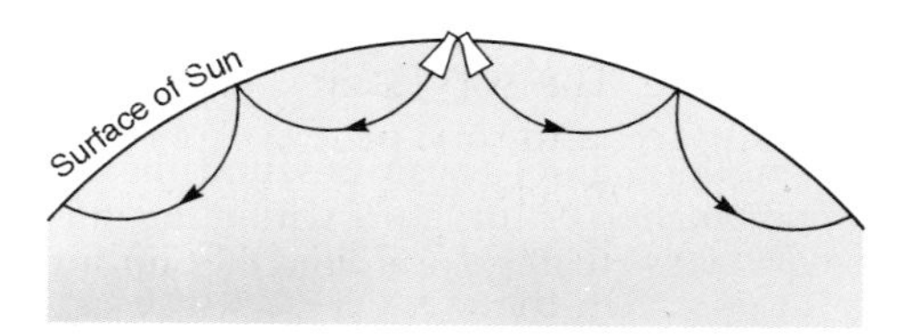

Figure B.4. If twin megaphones emit two beams of sound in opposite directions downwards into the Sun, then after one complete circuit, the two waves moving in opposite directions combine to form a pattern of standing waves

Now let us consider the waves emitted by our double megaphone towards the North and South, and the echoes that have completed one or more circuits of the Sun. We know that opposing wave-trains produce a pattern of standing waves. The trains of waves created by our double megaphone will also create a regular pattern of waves where the opposing waves reach the surface.

We have set up two megaphones at one specific spot, both pointing down at the same angle, and both emitting a note of the same pitch. In doing so, of course, we have grossly over-simplified the situation. In reality, *every* point on the Sun's surface emits sound, at *every possible* pitch, and it is radiated downwards in *all* directions. All these sound waves are bent back by the interior of the Sun and then reflected back into the interior by the surface. They run round the Sun in every direction simultaneously. The only waves that will remain in evidence at any one point, however, will be those that are reinforced by their echoes. All the others will cancel one another out, because the crests and troughs do not coincide with the returning crests and troughs.

In fact, it is the same as with Chladni's plates. Instead of a megaphone, it was a violin bow that started their vibration. The bending of the sound waves by the solar interior and their reflection by the surface corresponds to the reflection of the waves from the edges of Chladni's plates.

If the oscillations of the Sun are caused by sound waves, then we must expect a complicated pattern of oscillations to appear. I need to tackle another difficult concept here. We should not imagine that there are fixed nodal lines on the Sun, where the surface does not move. That would only be the case if there were just one spot on the Sun that emitted sound waves—i.e., just a single megaphone. No, every point on the Sun emits sound waves at every frequency and in every direction. The waves emitted from a specific point, which we may call P, reach every point in the body of the Sun. There are contributions from the waves coming from P in the motions of every other point of the Sun.

Although the image of repeatedly reflected sound waves that combine with one another is enlightening, it may also lead to misunderstandings. We have imagined the angle and frequency of our double-megaphone to be so arranged that the two waves, propagating towards the North and South, are reinforced by their echoes. The Sun's pattern of oscillations is created by waves (and their echoes) criss-crossing one another. We should not conclude from this, however, that everywhere that two opposing waves encounter the surface, there is an antinode. The waves may cancel one another out at a particular point on the surface, and thus produce a node.

The number of reflections undergone by a wave as it travels round the Sun therefore has nothing to do with the number of nodes, and thus with the degree of the oscillation. On the contrary: sound waves that dive more steeply beneath the surface, and thus travel deeper, have a lower degree.

This is easy to see. In a standing wave the nodes of the oscillation are exactly half a wavelength apart. Figure B.5, *top*, shows a pair of oppositely moving sound waves near the surface. The perpendicular lines represent the nodal surfaces of each wave, seen from the side. Where they cut the surface we have a node. Figure B.5, *bottom*, shows waves of the same frequency and wavelength, but where the waves travel at a much steeper angle. The distance between the nodes at the surface has become greater. That means that there are fewer nodes, and that therefore the degree of the oscillation associated with the waves that penetrate more deeply is lower.

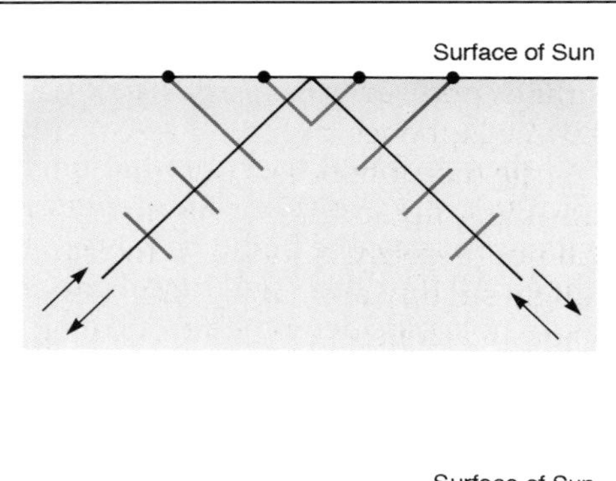

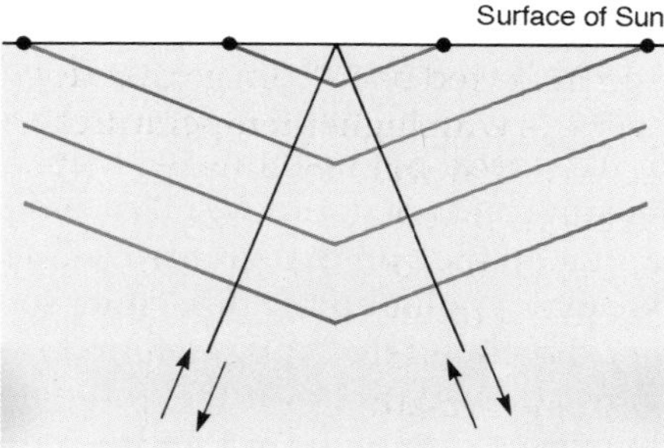

Figure B.5. Beams of sound directed at different angles to the surface. The wave-crests are indicated by the transverse lines. Between them, not marked, are the nodal surfaces of the standing-wave pattern that is produced. Top: The wave-trains descend at a shallow angle. The crests of the waves at the surface (solid dots) are close together, the degree of oscillation is high. Waves like these do not penetrate very deeply into the Sun's interior. Bottom: The wave-trains descend steeply. The wave-crests at the surface are far apart. The degree of oscillation is lower, and such waves penetrate more deeply (see Figure B.3)

Particularly significant is the extreme case of an oscillation with a degree of zero. The waves that are responsible for this oscillation travel down, perpendicular to the surface, towards the centre of the Sun, and are reflected back in the deeper layers. These waves penetrate more deeply than any others.

Temperature measurements by echo sounding

The image of beams of sound is an excellent means of envisaging how the pattern of oscillations at the surface provides information about the inner workings of the Sun. Let us reconsider our twin megaphone droning away with beams of sound directed North and South. If we adjust the frequency and angle so that a pattern of standing waves is created at the surface, then the beams of sound are reflected back to the surface from a specific depth. This may happen perhaps once, or perhaps many times, until the two wave-trains converge on the megaphones and fuse to create a standing wave. If we now alter the angle of the megaphones so that the signal penetrates slightly deeper, then after travelling up and down on their journey round the Sun, the sound-waves will not precisely combine at their point of origin. But we can again alter the angle slightly and, after a few attempts, find the setting at which the two echoes reach us together after having travelled round the Sun. We can now alter the frequency so that the two signals are reinforced by their echoes. We will again have a standing-wave pattern. If previously the two signals were reflected 10 times, let us say, within the interior, this may now happen 11 times. This is because the sound-waves from the more steeply angled megaphones are reflected back from greater depths (Figure B.3). They have passed through layers with higher temperatures and thus with a higher speed of sound, than those that were reflected just 10 times. This is apparent in the pattern of oscillations that they produce.

Anyone studying the Sun's oscillation patterns is able to determine something about the temperatures that prevail in its interior. The Sun's modes of oscillation do not provide just this information, however. We can, in principle, also determine something about the motions within its interior.

Echo-location and the rotation of the Sun's interior

As early as Scheiner's time, sunspots showed that the Sun rotates. Carrington determined the rotation period more accurately. Modern measurements making use of the Doppler effect have even shown that zones of the Sun's surface may temporarily rotate somewhat faster. What we know about the Sun's rotation, however, comes only from a study of the surface. We do not know, therefore, whether the interior rotates more slowly than the surface. Does it perhaps rotate faster?

The waves of the solar oscillation pass through the inner regions. Note that standing waves are produced by the superimposition of waves travelling in opposite directions that have the same frequency and strength. Let us now think of our twin megaphones again, which emit two beams of sound symmetrically downwards. This time we will imagine that we set them up on the equator, and that the beams are sent directly East and West. We adjust the angle and frequency until we have a standing-wave pattern. If the interior of the Sun rotates faster than the surface on which our megaphones are situated, then for one beam of sound the rotation deep inside the Sun acts as if it were a following wind, while for the other the rotation behaves as if it were a head wind (Figure B.6). This means that the pattern of waves on the surface is different from what it would be if the Sun rotated as a solid body. This is how, in principle, we may determine the rotation of the Sun's interior.

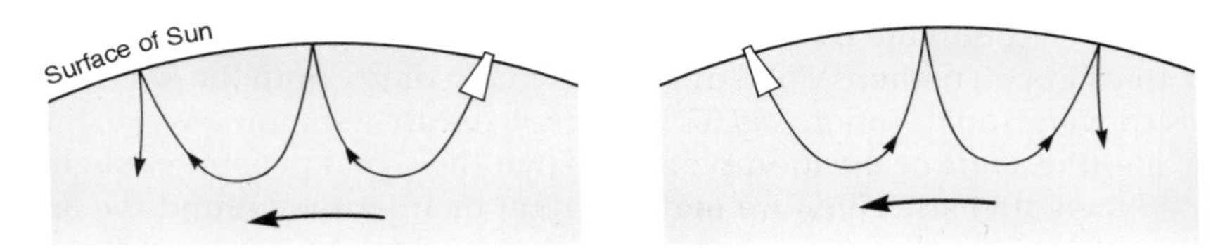

Figure B.6. The two beams of sound from a double megaphone, drawn separately. If the interior of the Sun rotates differently from the surface, with the relative motion being as shown by the slightly curved arrow, then the left-hand beam of sound experiences a following wind, whereas the right-hand one has the wind in its face. In the first case the speed of the wave-train will be increased, and in the second, decreased. The pattern of standing waves that the two create when they combine will be different from what it would be if the interior of the Sun rotated with the surface. From the pattern of standing waves, therefore, we are able to determine something about the rotation of the solar interior

C FREQUENCY AND WAVELENGTH

A wave, moving at the velocity of light, is shown in Figure C.1 at three different points in time. It begins (top) at one point, marked 'Start' and, one second later, reaches the 'Finish' as shown at the bottom. Because light travels 300 000 kilometres per second, the distance between the two points is 300 000 km. The frequency gives the number of wave-crests that cross into the region between 'Start' and 'Finish' in that one second. Because the distance between two crests defines the wavelength, it follows that the number of wave-crests times the wavelength must equal 300 000 km. From this we have: Frequency × Wavelength = Speed of light.

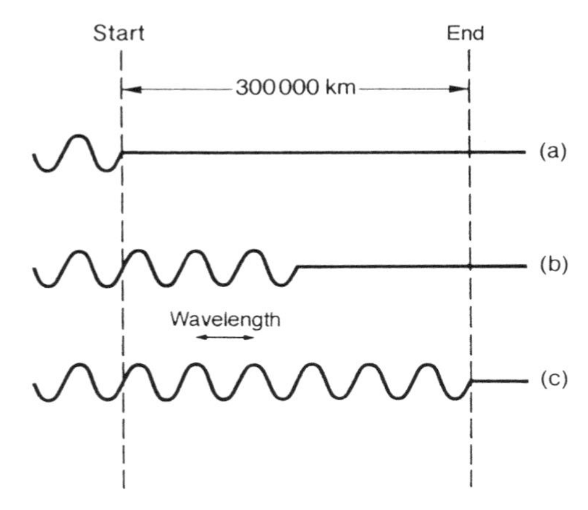

Figure C.1. A series of waves propagating at the speed of light begins at 'Start' and reaches 'End', 300 000 km away one second later

In the diagram the frequency is six oscillations per second, so the wavelength is therefore 50 000 km. If the wavelength were that of a typical medium-wave radio station, say 375 metres, then 801 000 wave-crests would travel past the starting position every second.

In the equation just given we may take the frequency as being measured in Hz, and the wavelength in metres. The velocity of light is 300 000 km per second. If we measure the frequency in MHz, then the frequency in Hz is one million times the value in MHz. Our equation then becomes:

$$1\,000\,000 \times \text{Frequency (in MHz)} \times \text{Wavelength (in metres)} = 300\,000\,000,$$

whence:

$$\text{Frequency (in MHz)} \times \text{Wavelength (in metres)} = 300.$$

INDEX